LIFE ON EARTH-LIKE PLANETS

Gaia and Her Sisters

LIFE ON EARTH-LIKE PLANETS

Gaia and Her Sisters

Gerald E Marsh, retired

Argonne National Laboratory, USA

World Scientific

NEW JERSEY · LONDON · SINGAPORE · BEIJING · SHANGHAI · HONG KONG · TAIPEI · CHENNAI · TOKYO

Published by

World Scientific Publishing Co. Pte. Ltd.

5 Toh Tuck Link, Singapore 596224

USA office: 27 Warren Street, Suite 401-402, Hackensack, NJ 07601

UK office: 57 Shelton Street, Covent Garden, London WC2H 9HE

Library of Congress Control Number: 2024944761

British Library Cataloguing-in-Publication Data
A catalogue record for this book is available from the British Library.

LIFE ON EARTH-LIKE PLANETS
Gaia and Her Sisters

ISBN 978-981-12-9507-2 (hardcover)
ISBN 978-981-12-9508-9 (ebook for institutions)
ISBN 978-981-12-9509-6 (ebook for individuals)

For any available supplementary material, please visit
https://www.worldscientific.com/worldscibooks/10.1142/13897#t=suppl

Desk Editors: Yong Qi Soh/Muhammad Ihsan Putra

Typeset by Stallion Press
Email: enquiries@stallionpress.com

To Marlena Ekstein for her friendship and help during very difficult times.

Preface

Gaia was the ancient Greek goddess that personified the Earth, but ever since the publication of the 1979 book *Gaia* by J.E. Lovelock for many it has come to mean a single complex living "organism" where the interaction between the organic and inorganic parts of the planet keeps Earth a fit place for life. One thing for sure, since the Hadean period of Earth's existence, covering the period from about 4.3 to 3.8 billion years ago, life has massively transformed the planet thereby allowing life as we know it today to exist.

Unfortunately, we have only one Earth-like planet that can be used to extrapolate to others, our own. But, as will be shown, since the development of life on Earth was not a stochastic process, it is likely to nonetheless be an excellent example for the development of life on most if not all Earth-like planets. Of course, this does not mean that the life forms that evolve will be identical to those on Earth, only that they must satisfy the same constraints. What these are is the primary subject of this book.

During the Hadean period of the Earth's existence newly formed oceans covered a greater portion of the Earth's surface than today. At that time there was little oxygen in the atmosphere so that there would have been no ozone layer to protect the Earth from the harsh ultraviolet radiation from the sun.[1] Life, when it emerged about

---

[1]I should mention that there is something called the Faint Young Sun problem, a contradiction between stellar evolution models that predict a ~25% lower energy input to the Earth's climate system during the Archean period some 3.8–2.5 billion years ago than there is today, while there is ample evidence for the presence of liquid surface water and life during the Archean. Therefore, there must have been

3.8 billion years ago, or a little after, would have had to have been in the ocean. This was at a time when the moon was much closer to the Earth, creating very high tides, and when the day was much shorter. The processes leading to life would have had to have been thermodynamically favorable under these conditions or life would not have arisen at all. What this means is discussed in Chapter 1.

It is an obvious statement that the appearance of life is ultimately founded on the ability of matter, governed by the principles of quantum mechanics, to form the molecules need for life to exist. These molecules combine, overcoming the limits normally imposed by both unfavorable free-energy constraints and activation energies, because of the properties of driven nonlinear thermodynamic systems. Most stars have planets, and those with Earth-like planets are all very likely to have life due to the ability of matter to form complex biomolecules. Complex biomolecules are also found in meteorites and gas clouds in space.

Chapter 2 explains the statistical approach to the origin of life, and Chapters 3, 4 and 5 explain why the statistical approach to estimating the probability of life on Earth-like planets is invalid. Specifically, Chapter 3 covers abiogenesis, Chapter 4 the prokaryote–eukaryote divide, and Chapter 5 the evolution of polymers from monomers found in the prebiotic environment. Chapter 6 covers the evolution of multicellular life and its symmetries, and Chapter 7 gives some necessary background for the following chapters.

In this book, life is treated as a natural process governed by the same laws as nonliving processes; it is the inevitable outcome of biochemical forces woven into the fabric of the universe. The evolution toward increasing complexity is also a natural process, which would appear to violate the second law of thermodynamics, but it does not because of other compensating factors. The second law tells us that entropy can only increase or remain constant in a closed, isolated systems where no heat can enter or leave. Life does not satisfy this requirement because it can exchange both matter and energy with its environment so that entropy can decrease in a living

---

a compensating effect or–despite the best efforts of many people–the models are wrong. (See, Feulner, G., "The Faint Youn Sun Problem", *Rev. Geophys.* **50**, (2012)).

system in a way fully compatible with the second law. Life is an open system; its anti-entropic nature being driven mostly by solar energy.

The appendices cover some material and concepts that supplement the subject matter of some of the chapters. Appendix 1 is a limited introduction to the Bayesian and frequentist interpretation of probability as well as to Markov processes; Appendix 2 is a basic introduction to metabolism; and Appendix 3 discusses the hearing in humans and dolphins, which have a high intelligence and will serve as an example when considering the possibility of high-level intelligence on Earth-like planets. Appendix 4 is an introduction to symbiosis; Appendix 5 discusses the implications of the possibility of intelligent life on Earth-like planets in the context of the relationship between science and religion; and Appendix 6 discusses emergent behavior in the biological sciences.

Intelligent life on Earth-like planets, and the requirements for it to arise, is the subject of Part II of this book. Our Earth serves as a compelling example for what is required for Earth-like planets to evolve intelligent life and especially the high-level intelligence found in modern humans.

The human cerebral cortex is unique not only because of the number of neurons it contains but because it doubled in size over the unusually short period of the last 1.5 million years. This is the source of the higher-level intelligence found in modern humans. Understanding how this happened on Earth is crucial for trying to ascertain whether such intelligence could arise on other Earth-like planets. This is the subject of Chapter 8. Chapters 9, 10, and 11 cover the requirements for higher level intelligence to arise on Earth-like planets given the example of our own Earth.

Chapter 9 introduces the fact that the human brain is not the only one on Earth that has undergone a significant increase in size and complexity. Although they have taken a very different anatomical path to neural complexity, Cetaceans may be the best example we have of an alternate route to intelligence comparable–although different–to that of humans. Understanding, as best as we can, the high level of intelligence found in dolphins is a worthwhile exercise in the context of trying to explore the possibility of high-level intelligence on Earth-like planets.

Contents

Part I
Abiogenesis on Earth-Like Planets

Chapter 1

Nonequilibrium Thermodynamics and the Origin of Life[*]

As stated in the Preface, the processes leading to life on the early Earth would have had to have been thermodynamically favorable. This chapter explains why this was indeed the case. The contents of the chapter apply to all Earth-like planets as well as to Earth.

It is generally believed that life began on Earth with the evolution of self-replicating polynucleotides. Ever since the paper by Gilbert[1] that discussed the possibility that catalytic RNA enzymes or ribozymes could be involved in the evolution of life, and incidentally coined the term the "RNA world", RNA has been the favorite molecule. Gilbert's *News and Views* paper addressed the previous week's *News and Views* paper by Westheimer.[2] The original discovery of the enzymatic activities of RNA was by Cech.[3] In 2009, Lincoln and Joyce[4] showed the self-sustained replication of two ribozymes that catalyze each other's synthesis. These cross-replicating ribozymes grew exponentially in the absence of proteins or other biological materials.

---

[*]Marsh, G.E., "Thermodynamics and the Origin of Life", *Canadian J. Phys.*, **100** (2022), 285–291.

[1]Gilbert, W., "The RNA world", *Nature* **319** (20 February 1986), 618.

[2]Westheimer, F.H., "Biochemistry: Polyribonucleic acids as enzymes", *Nature* **319**, (13 February 1986), 534–536.

[3]Cech, T.R., "Self-splicing RNA: Implications for evolution", *Int. Rev. Cytol.* **93** (1985), 3–22.

[4]Lincoln, T.A. and Joyce, G.F, "Self-sustained replication of an RNA enzyme", *Science* **323** (27 February 2009), 1229–1232.

RNA molecules are composed of purine and pyrimidine nucleotides. How these nucleotides could have formed concurrently under the geophysical constraints of the early Earth was an unsolved chemical mystery before Becker *et al.*[5] published their paper "Unified prebiotic plausible synthesis of pyrimidine and purine RNA ribonucleotides" in 2019. They found a reaction network under which both nucleotides could simultaneously form, being driven by wet-dry cycles. The chemical reactions involved are very complex, but Hud and Fialho[6] have summarized them as shown in Figure 1.

The geochemical conditions in the current work of Becker *et al.* is compatible with their previous synthesis of the purine nucleosides.[7] The wet-dry cycles allow the dried reactants to coalesce into a

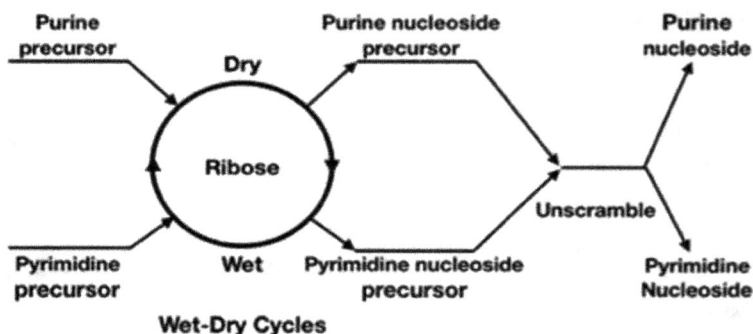

Figure 1. A simplified reaction network for the simultaneous formation of purine and pyrimidine nucleosides. When a mixture of the sugar ribose and nonnatural nucleobases are subjected to wet-dry cycles, intermediate molecules are formed that can be converted to the natural nucleosides, which when phosphorylated give the natural nucleotides. For a discussion of what "scrambled" means see the article by Hud and Fialho from which this figure is adapted. They used the term to designate the precursor formed by Becker *et al.*, which they called a "scrambled" pyrimidine nucleobase.

---

[5]Becker, S. *et al.*, "Unified prebiotically plausible synthesis of pyrimidine and purine RNA ribonucleotides", *Science* **366** (4 October 2019), 76–82.

[6]Hud, N.V. and Fialho, D.M., "RNA nucleosides built in one prebiotic pot", *Science* **366** (4 October 2019), 3–33.

[7]Becker, S. *et al.*, "A high-yielding, strictly regioselective prebiotic purine nucleoside formation pathway", *Science* **352** (13 May 2016), 833–836.

concentrated state where their joining through the covalent bond formation with the release of water molecules becomes thermodynamically favorable. What "thermodynamically favorable" means is the subject of much of what follows in this chapter.

From a simplistic point of view, life appears to violate the second law of thermodynamics since ordered life forms have a lower entropy than their precursors. Of course, this apparent violation vanishes when one realizes that life processes draw upon the free energy from the surroundings so that total entropy increases. In terms of the origin of life, it has been known for some time that matter can spontaneously organize itself if the Gibbs free energy for the process is negative or if an external source of "activation energy" is provided. If this absorbed energy is dissipated after overcoming the activation barrier it is no longer available to drive the reverse process. This irreversible phenomenon is called "driven self-assembly". Explaining how this works for a macroscopic system requires the introduction of a generalized form of the second law of thermodynamics.

Nonequilibrium States

A system may be maintained in a nonequilibrium state by a flow of energy. If the state is time-independent, macroscopic observables will have constant nonequilibrium values. The example often given is a constant electrical current through a resistor with a steady rate of heat generation. Such dissipative structures can be formed and maintained by irreversible processes that continuously increase entropy.[8],[9] In a linear regime, small deviations in the forces driving a flow (such as the heat flow on the resistor example) will lead to the flow being a linear function of the forces driving the flow. But a system that is not in thermodynamic equilibrium need not be in a stationary, time-independent state since systems that are far from

---

[8]Kondepudi, D. and Prigogine, I., *Modern Thermodynamics: From Heat Engines to Dissipative Structures* (John Wiley & Sons, Ltd., United Kingdom, 2015).

[9]DeBari, B. *et al.*, "Oscillatory dynamics of an electrically driven dissipative structure", *PLOS ONE* (May 29, 2019). https://doi.org/journal.pone.0217305.

equilibrium can become dependent on nonlinear phenomenological laws.

Systems of identical noninteracting components in thermodynamic equilibrium can be described by the Boltzmann distribution (also known as the Gibbs distribution), which gives the equilibrium probability distribution of different energy states of a system as a function of the state's energy and the temperature of the system. It has the general form $P_i \propto \exp[-E_i/k_BT]$, where k_B is the Boltzmann constant, P_i is the probability that the system is in the state i and E_i is the energy of the state. Thus, if the states i and j have energies E_i and E_j, the relative probability is

$$\frac{P_i}{P_j} = \exp\left[\frac{E_j - E_i}{k_BT}\right]. \tag{1}$$

Moreover, systems in thermodynamic equilibrium satisfy the principle of detailed balance, which is equivalent to microscopic reversibility.

If the system is driven far from equilibrium into the nonlinear regime, the Boltzmann distribution is no longer valid since the thermodynamic flows are no longer linear functions of the thermodynamic forces. Far from equilibrium states can evolve into one of many new, highly organized states known as dissipative structures.

Microstates and Reversibility

The statistical behavior of nonequilibrium systems requires the introduction of comparisons between the dynamical trajectories of the components of the system rather than the local properties of individual microstates at one moment of time.[10] If the dynamics of a system are stochastic and Markovian (meaning a sequence of events where the probability of each event depends only on the state of the previous event), one can require that the dynamics follow the

---

[10]English, J.L., "Dissipative adaptation in driven self-assembly", *Nat. Nanotechnol.* **10** (November 2015), 919–923.

microscopically reversible condition[11]

$$\frac{P[x(+t)|\lambda(+t)]}{P[\bar{x}(-t)|\bar{\lambda}(-t)]} = \exp\{-\beta Q[x(+t), \lambda(+t)]\}, \tag{2}$$

where $\beta = 1/k_B T$, the state of the system is given by the function x, representing all dynamical uncontrolled degrees of freedom, while λ is a controlled time-dependent parameter.

In Equation (2), $P[x(+t)|\lambda(+t)]$ is the probability of following the path $x(+t)$ through phase space and the denominator is the corresponding time-reversed path. This notation for the time-reversed path is a consequence of changing the time origin so that $t \in \{-\tau, \tau\}$, where τ could be infinite. The overbar indicates that quantities odd under time reversal also change sign. Q is the amount of energy in the form of heat transferred to the system from the heat bath. Q is a function of the phase space path and odd under time reversal; i.e., $Q[x(+t), \lambda(+t)] = -Q[\bar{x}(-t), \bar{\lambda}(-t)]$.

Equation (2) is microscopically reversible; it relates the probability of a given path to its reverse path. Note that this is not the same as the principle of detailed balance, which refers to the probabilities of changing states independent of path. This distinction is important because Equation (2) holds when the system is driven by an external time-varying force field. English[12] has used this equation to derive a generalization of the second law of thermodynamics that is important for many far-from-equilibrium thermodynamic systems. It applies to the macroscopic transition between complex course-grained states.

Dissipative Adaptation

To be consistent with English's notation, rewrite Equation (2) as

$$\frac{\pi[\gamma]}{\pi^*(\gamma^*)} = \exp\left[\frac{\Delta Q(\gamma)}{kT}\right], \tag{3}$$

[11] Crooks, G.E., "Entropy production fluctuation theorem and the nonequilibrium work relation for free energy differences", *Phys. Rev. E* **60** (1999), 2721–2726.
[12] English, J.L., "Statistical physics of self-replication", *J. Chem. Phys.* **139** (2013), 121923–121928.

where γ is a microtrajectory and the "*" indicates time-reversed. Here, it is assumed that $\Delta Q(\gamma)$ is the sum of the internal energy change of the system when traversing the path γ and the work applied to the system by an external field; i.e., $\Delta Q(\gamma) = \Delta E + W$.

Let i, j, and k represent different possible configurations of a system composed of distinct components (such as particles). Equation (3) then becomes

$$\frac{\pi[i \to j]}{\pi^*(j^* \to i^*)} = \exp\left[\frac{\Delta Q(i \to j)}{kT}\right]. \tag{4}$$

Using Equation (4) also for the trajectory $i \to k$ and taking the ratio of the equations for the transition $i \to j$ and $i \to k$, that is,

$$\frac{\frac{\pi[i \to j]}{\pi^*(j^* \to i^*)}}{\frac{\pi[i \to k]}{\pi^*(k^* \to i^*)}} = \frac{\exp\left[\frac{\Delta Q(i \to j)}{kT}\right]}{\exp\left[\frac{\Delta Q(i \to k)}{kT}\right]}, \tag{5}$$

with the substitution $\Delta Q(\gamma) = \Delta E + W$, after some algebra Equation (5) will yield

$$\frac{\pi[i \to j]}{\pi[i \to k]} = \exp\left[\frac{-\Delta E(j \to k)}{kT}\right] \frac{\pi^*(j^* \to i^*)}{\pi^*(k^* \to i^*)} \frac{\exp\left[-\frac{W(i \to k)}{kT}\right]}{\exp\left[-\frac{W(i \to j)}{kT}\right]}. \tag{6}$$

The first term on the right-hand-side of Equation (6) comes from the term $\exp\left[\frac{\Delta E(i \to j) - \Delta E(i \to k)}{kT}\right]$ encountered when doing the algebra. The energy level relationships of Equation (6) are shown in Figure 2.

From this figure, one can see that $\Delta E(i \to j) - \Delta E(i \to k) = \Delta E(j \to k)$.

To explain the concept of dissipative adaptation, English assumes that the states have the same energy so that the first term on the right-hand side of Equation (6) is unity, and averages over

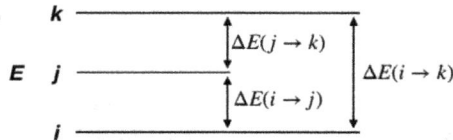

Figure 2. Energy level relationships of Equation (6).

all microtrajectories with fixed endpoints. With these assumptions, Equation (6) becomes

$$\frac{\pi[i \to j]}{\pi[i \to k]} = \frac{\pi^*(j^* \to i^*)}{\pi^*(k^* \to i^*)} \frac{\exp\langle -\frac{W(i \to k)}{kT} \rangle}{\exp\langle -\frac{W(i \to j)}{kT} \rangle}. \tag{7}$$

The brackets $\langle \ldots \rangle$ indicate the average over microtrajectories. Note that even though the states now have the same energy, the left-hand side of Equation (7) could differ from unity since not all states of the system are equally accessible in a finite time.

For different oscillatory external forces, known as the "drive", different configurations of the system will absorb work from the external forces at different rates. A given system configuration could then surmount "activation barriers" to transition to states that would not be accessible through thermal fluctuations alone. This is shown in Figure 3 for the general case where the energies of the states are not identical.

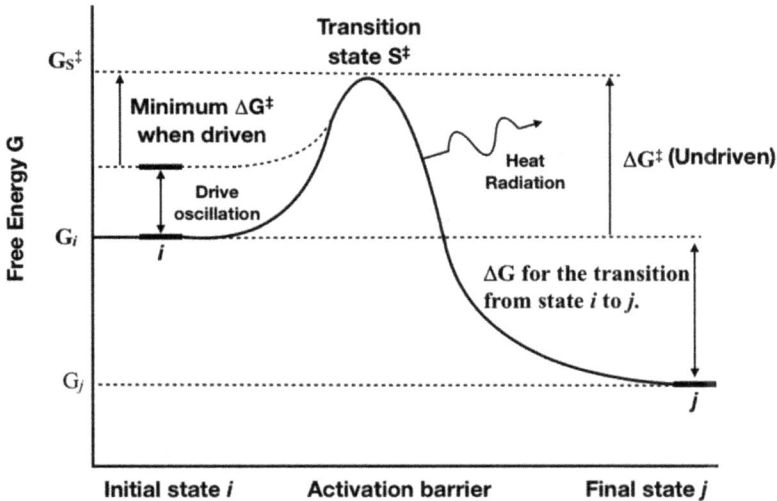

Figure 3. Transition through a system configuration activation barrier. $\Delta G^\ddagger = G_S^\ddagger - G_i$, where G_i oscillates when driven. For the case where the energies of the states i and j are equal, $G_i = G_j$. The energy absorbed from the drive is radiated as heat in the transition from $S^\ddagger$ to state j. This figure can be compared to Fig. A.1 in the appendix for the activation energy of a chemical reaction.

The point of all this is that some of the randomly changing system configurations will be better able to absorb work from the drive than others and this leads to a mostly one-directional change in configurations — due to the loss of the radiated heat energy — to those better able to absorb and dissipate the energy absorbed from the drive. As put by English, the structure will appear to self-organize into a state that is well adapted to the environmental conditions set by the drive. This, he calls the phenomenon of "dissipative adaptation".

In this vein, Kondepudi and Prigogine in their 2015 book *Modern Thermodynamics*, referenced earlier, introduce the concept of "structural instability" that can occur when a new chemical species is introduced into a nonequilibrium chemical system that could destabilize the system so that it evolves into a new state. They liken this to Darwinian evolution at the molecular level. Thus, in their words "we see instability, fluctuation and evolution to organized states as a general nonequilibrium process whose most spectacular manifestation is the evolution of life."

Macroscopic Irreversibility

The relationship between microscopic irreversibility and entropy production has thermodynamic effects on far-from-equilibrium macroscopic processes and in particular on biological self-replication (to be discussed in the next section). English derived a generalized form of the second law of thermodynamics that can be written as

$$\beta \langle \Delta Q \rangle_{I \to II} + \ln \left[\frac{\pi(II \to I)}{\pi(I \to II)} \right] + \Delta S_{\text{int}} \geq 0. \tag{8}$$

The derivation of this equation begins with Equation (3) with $\gamma \to x(t)$, where $0 \leq t \leq \tau$. Taking the natural logarithm of the resulting equation gives

$$\beta \Delta Q = \ln \left[\frac{\pi[x(t)]}{\pi[x(\tau - t)]} \right], \tag{9}$$

an equation concerned with microscopic irreversibility.

Suppose there is a coarse-grained observable, I, which can be associated with a probability distribution $p(i|I)$, the probability that it is in a microstate i. If there is a second course-grained observation of the system after a time interval τ designated by II, $p(j|II)$ is defined as the probability that the macrostate II (that evolved from macrostate I after a period of time τ) is in the microstate j. Note that the macrostates I and II are ensembles of paths.

Crucially, these probability functions allow a macroscopic definition of irreversibility:

$$\pi(I \to II) = \int_{II} dj \int_{I} di\, p(i|I)\pi(i \to j),$$

$$\pi(II \to I) = \int_{I} di \int_{II} dj\, p(j|II)\pi(j \to i). \tag{10}$$

Similar to what was done in Equation (5), taking the ratio of $\frac{\pi(II \to I)}{\pi(I \to II)}$ gives, after some algebra,

$$\frac{\pi(II \to I)}{\pi(I \to II)} = \langle\langle e^{-\beta \Delta Q_{ij}} \rangle_{i \to j} / \exp\left[\ln\left[\frac{p(i|I)}{p(j|II)}\right]\right]\rangle_{I \to II}. \tag{11}$$

The first averaging bracket on the right-hand side is the average of all paths from $i \in I$ to $j \in II$, each path being weighted by its likelihood (the second averaging bracket).

The next step is to introduce the Shannon entropy $S = -\sum_i p_i \ln p_i$ so that an expression for the internal entropy change for the transition between the ensembles $I \to II$ can be written. English uses units such that the Boltzmann constant is unity. This results in $\Delta S_{\text{int}} = S_{II} - S_I$ and after some additional algebra he obtains the generalization of the second law of thermodynamics given by Equation (8); that is,

$$\beta \langle \Delta Q \rangle_{I \to II} + \ln\left[\frac{\pi(II \to I)}{\pi(I \to II)}\right] + \Delta S_{\text{int}} \geq 0. \tag{12}$$

The first term in Equation (12) is the entropy change of the heat bath and the entropy generated by the second term vanishes if $\pi(II \to I) = \pi(I \to II)$, which would result in the usual second law of thermodynamics where $\beta \langle \Delta Q \rangle_{I \to II} + \Delta S_{\text{int}} \geq 0$ because the average entropy change of the universe must be greater than or equal to zero.

One might question whether the use of the Shannon entropy is legitimate. The Boltzmann distribution implies the usual thermodynamic definition of entropy. The Gibbs–Shannon entropy given by

$S = -k_B \sum_i p_i \ln p_i$ is equivalent to the thermodynamic definition of entropy only for what is known as the generalized Boltzmann distribution,[13] which is valid for all Markovian systems even those not in thermodynamic equilibrium. The generalized Boltzmann distribution itself was defined by Lin[14] using an analogy based on electronics to give an explanation of the concept. Given a thermodynamic system with $m + n$ generalized forces and coordinates, Xiang *et al.* write the probability density function $\Pr(\vec{\omega})$ of the microstate $\vec{\omega}$ for the generalized Boltzmann distribution as

$$\Pr(\vec{\omega}) \propto \exp\left[\sum_{\eta=1}^{n} \frac{X_\eta x_\eta^{(\vec{\omega})}}{k_B T} - \frac{E^{(\vec{\omega})}}{k_B T}\right], \tag{13}$$

where the X_η are generalized forces and E and x_η are random variables, and Xiang *et al.* use the vector notation to designate a microstate, as in $\vec{\omega}$. If the generalized forces vanish, Equation (13) reduces to the usual Boltzmann distribution.

Since the systems considered above are Markovian, the use of the Shannon entropy is indeed legitimate.

Replicating Systems and the Generalized Second Law of Thermodynamics

English has used self-replicating systems to illustrate the use of the generalized second law. Consider $n(t = 0) \gg 1$ for n self-replicating molecules at an inverse temperature β. They would have an exponential growth given by

$$n(t) = n(0)e^{(g-\delta)t}, \tag{14}$$

where g determines the growth rate and δ the decay rate.

The probability that in a time dt one particular replicator associated with $\pi(I \to II)$ would replicate would be given by gdt

---

[13]Gao, Xiang *et al.*, "The generalized Boltzmann distribution is the only distribution in which the Gibbs–Shannon entropy equals the thermodynamic entropy", (12 April 2019). arXiv:1903.02121 [cond-mat.stat-mech].

[14]Lin, M.M., "Generalized Boltzmann distribution for systems out of equilibrium", (13 October 2018). arXiv:1610.02612 [cond-mat.stat-mech].

and its decay probability $\pi(II \to I)$ would be $\delta \, dt$. Imposing the generalized second law of thermodynamics yields,

$$\Delta S_{\text{tot}} = \beta \Delta q + \Delta S_{\text{int}} \geq \ln \left[\frac{g}{\delta} \right]. \tag{15}$$

Note that if $g > \delta$, so that there is net growth, the total entropy associated with self-replication will have a positive lower bound.

Two conclusions are readily apparent: (1) The growth rate of a self-replicator depends on its internal entropy (ΔS_{int}), its durability ($1/\delta$), and the heat (Δq) dissipated into the surrounding heat bath in the process of replication; and (2), heat must be generated from energy stored in the reactants or work done on the system by a time-varying external driving field.

Implications for the Origin of Life

Using the RNA molecule again as an example, how purine and pyrimidine nucleosides could have formed together under early Earth geophysical constraints was, as mentioned in the Introduction, until recently an unsolved chemical problem. From a global perspective, the resolution of this problem involved far from equilibrium thermodynamics — in this case where the system was driven by wet-dry cycles; it is this external forcing that made the formation of these nucleosides thermodynamically favorable.

While the idea that far from equilibrium thermodynamics is fundamental to the origin of life remains somewhat controversial, the case is quite strong for this point of view. There is also a simulated toy chemical model whose behavior is consistent with the idea of far-from-equilibrium self-organization.[15]

General chemistry tells us that the free energy change in a reaction is governed by the equilibrium constant. Furthermore, while the reaction rate can be changed by enzymes which do not alter the equilibrium, the change in free energy is independent of the path or the molecular mechanism of the transformation. It is clear that

---

[15]Kachman, T., Owen, J.A. and England, J.L., "Self-organized resonance during search of a diverse chemical space", *Phys. Rev. Lett.* **119** (2017), 038001-1–038001-5.

this is not the case for nonequilibrium thermodynamics "whose most spectacular manifestation is the evolution of life".[16]

Our Earth formed some 4.5 billion years ago and there is strong evidence that life appeared 700 million to 1 billion years later; that is, about 3.8 billion years ago. Far from equilibrium nonlinear thermodynamics in the presence of external drives could help explain this very rapid origin of life. Because the length of the day was only around seven or so hours long 3.8 billion years ago, the tides would be far higher than today and there would be strong diurnal forces that could play the role of an external drive including UV radiation, which would be far more intense than today since there was little if any oxygen in the atmosphere and therefore no ozone layer to block the UV.

Life appears as a process ultimately founded on the ability of matter, governed by the principles of quantum mechanics, to form the molecules need for life to exist. These molecules combine, overcoming the limits normally imposed by both unfavorable free energy constraints and activation energies, because of the properties of driven nonlinear thermodynamic systems.

Most stars have planets, and those with Earth-like planets are all very likely to have life due to the reduction in the activation energy for the formation of complex biomolecules arising from driven, far-from-equilibrium nonlinear thermodynamics.

---

[16]Kondepudi, D. and Prigogine, I., *Modern Thermodynamics: From Heat Engines to Dissipative Structures* (John Wiley & Sons, Ltd., United Kingdom, 2015).

Chapter 2

Statistical Approach to the Origin of Life

Ever since it became known in the 17th century that the stars were like our own sun and could have their own planets,[1] an idea popularized by Bernard de Fontenelle's 1686 book *Entretriens sur la pluralite des mondes*, people wondered if life could exist on these planets. But the question is now generally phrased as "What is the chance, or probability, that life exists on these planets?". The very phrasing of the question in this way subsumes that probability theory is a legitimate means to arrive at an estimate of the possibility of life. Now that it is known that many stars have Earth-like planets, the question becomes far more important.

Estimating the probability of life on Earth-like exoplanets goes back to the Drake equation, which is a product of several probabilistic terms. Drake has said recently that "It has been gratifying that over the years the Equation has not been found erroneous — it is alive and well in its original form."[2] But it is also worth quoting Wikipedia on the subject:

> "Criticism related to the Drake equation focuses not on the equation itself, but on the fact that the estimated values for several of its factors are highly conjectural, the combined multiplicative effect being that the uncertainty associated with any derived value is so large that the equation cannot be used to draw firm conclusions."

---

[1]Even earlier, in 1584, the Dominican Friar Giordano Bruno published the book *De l'infinito universon e mondi* where he proposed that the stars were suns and Earth was one of many inhabited worlds in an infinite universe.

[2]Drake, F. *Int. J. Astrobiol.* **12**, (2013), 173–176.

The Drake equation was originally introduced by Drake to estimate the likelihood of detecting radio signals from advanced extraterrestrial intelligences. Nonetheless, the Drake equation is the origin of past and recent attempts to approach the problem of the origin of life from a statistical point of view.[3]

The use of statistics to estimate the probability of life means that its origin is considered to be a random or stochastic process where it is assumed that the origin is associated with a random probability distribution, of which there are different types.

More recent work in the statistical approach uses a Bayesian statistical framework and includes the work by Spiegel and Turner[4] who conclude that

> "Although terrestrial life's early emergence provides evidence that life might be abundant in the universe if early Earth-like conditions are common, the evidence is inconclusive and indeed is consistent with an arbitrarily low intrinsic probability of abiogenesis."

A later work by Kipping[5] builds on what they call the seminal paper by Spiegel and Turner, which applied a Bayesian formalism for interpreting the early emergence of life on Earth. Kipping's paper concludes that

> "... our analysis purely concerns the Earth, treating abiogenesis as a stochastic process against a backdrop of events and conditions which might be plausibly unique to Earth. If conditions sufficiently similar to the early conditions exist and sustain on other worlds for 1 Gy or more, then our analysis would then favor the hypothesis that life is common ..."

A discussion of the different statistical formulations is given in Appendix A.

Spiegel and Turner use a simplistic Poisson or uniform rate model for the origin of life. This means that they assume a Poisson process

---

[3]Lineweaver, C. and Davis, T.M., "Does the Rapid appearance of life on earth suggest that life is common in the universe", *Astrobiology* **2**, (2002), 293–304.
[4]Spiegel, D.S. and Turner, E.L., *PNAS* **109** (2012), 395–400.
[5]Kipping, D., *PNAS* **117** (2020), 11995–12003.

for a fixed period of time where Earth is lifeless before the period and cannot develop life after this period of time. They admit that their formulation has serious problems, the most important being that

> "... it treats abiogenesis as though it were a single instantaneous event and implicitly assumes that it can occur in only a single way ...and only in one type of physical environment. It is, of course, far more plausible that abiogenesis is actually the result of a complex chain of events that take place over some substantial period of time and perhaps via different pathways and in different environments."

Assuming that the "chain of events" is such that each step only depends on the current status, the chain of events is reminiscent of a Markov process, which is also discussed in Appendix A.

A 2021 paper by Snyder-Beattie *et al.*,[6] addressing the rise of intelligent life, concludes that

> "It took approximately 4.5 billion years for a series of evolutionary transitions resulting in intelligent life to unfold on Earth. In another billion years, the increasing luminosity of the Sun will make Earth uninhabitable for complex life. Intelligence therefore emerged late in Earth's lifetime. Together with the dispersed timing of key evolutionary transitions and plausible priors, one can conclude that the expected transition times likely exceed the lifetime of Earth, perhaps by many orders of magnitude."

This paper argues that eukaryogenesis is one of the key evolutionary transitions that could have been extraordinarily improbable and also uses a Bayesian statistical framework to draw its conclusions.

The probability of abiogenesis was first formalized in a Bayesian framework by Carter[7] who argued that the selection effect of our existence on Earth where abiogenesis occurred implies that nothing can be concluded about the probability of abiogenesis on Earth-like

---

[6]Snyder-Beattie, A.E. *et al.*, "The timing of evolutionary transitions suggests intelligent life is rare", *Astrobiology* **21** (2021), 265–278.

[7]Carter, B., "The anthropic principle and its implications for biological evolution", *Phil. Trans. Roy. Soc.* **A310** (1983), 347–363.

planets. A refutation of how the Bayesian approach had been used has been given by Whitmire[8] who argues that

> "the Carter conclusion is based on what is known as the 'Old Evidence Problem' in Bayesian Confirmation Theory and that taking this into account, the observation of life on Earth is not neutral but evidence that abiogenesis on Earth-like planets is *relatively* easy."

He concludes through a different argument that "the occurrence of abiogenesis on Earth-like planets is not rare."

If, as claimed in the Preface, life does appear as a process ultimately founded on the ability of matter, governed by the principles of quantum mechanics, to form the molecules need for life to exist, the statistical approach to finding the probability of life appearing on Earth-like planets is not valid because the ability of matter to form the molecules of life has the nature of a physical law. Physical laws are not statistical in nature, only sets of measurements related to them would be statistical. Abiogenesis, and the rise of eukaryotes or intelligence, is not a stochastic process.

Not being a stochastic process does not mean that the form life takes occurs by physical law alone. While the structure and variety of all atoms are determined by the rules of quantum mechanics, this does not limit the form of the lattice they or their compounds form, which depends on emergent degrees of freedom such as temperature and pressure reflecting environmental factors (Appendix F gives a discussion of emergent behavior). Consider the chemistry of saturated hydrocarbons. The rules of quantum mechanics certainly determine the bonding of carbon and hydrogen, and no matter how structurally complex the hydrocarbon, these rules are faithfully obeyed. But the rules of quantum mechanics say nothing about how many carbon atoms may form a chain or whether they form straight chains or branched-chain carbon skeletons. There are emergent degrees of freedom that appear when atoms combine to form these hydrocarbon molecules. Their structure is not fully determined by

---

[8]Whitmire, D.P., "Abiogenesis: The Carter argument reconsidered", *Int. J. Astrobiol.* **22** (2022), 94–99.

the underlying quantum mechanical rules governing the bonding of hydrogen and carbon atoms but rather by environmental influences. But, none the less, none of this is stochastic in nature.

The point of view taken here is that the emergence of life is a result of the normal working of the universal physical laws that govern the formation of biomolecules and ultimately life itself. There is no reason to think that they would not apply to life on other Earth-like planets as well.

Another Example of the Inappropriate Use of a Probabilistic Argument

Attempts have also been used to apply statistical considerations to what I have called The Problem of the "Prebiotic and Never Born Proteins".[9] The referenced paper addresses the argument that the limited set of proteins used by life as we know it could not have arisen by the process of Darwinian selection from all possible proteins. This probabilistic argument has a number of implicit assumptions that are not warranted. As will be seen, the number of amino-acid sequences that need have been sampled during the evolution of proteins is far smaller than assumed by the argument.

The genetic code specifies 20 amino acids, and a typical protein might be made up of a sequence of 200 amino acids. The argument is often made that the probability is negligible that the relatively limited set of something like 20,000 proteins coded for in the human genome (or the far greater number found in the natural proteome) arose by Darwinian selection of random variations out of the 20^{200} possibilities.[10] As put by Chiarabelli and Lucrezia,

> "Nature could not have explored all possible amino acid combinations and, therefore, that many proteins with interesting new properties could have never been sampled by Nature. By and large,

---

[9]Marsh, G.E., *Int. J. Astrobiol.* **12**, (2013).

[10]Chiarabelli, C. and De Lucrezia, D., "Question 3: The worlds of the prebiotic and never born proteins", *Orig. Life Evol. Bioph.* **37** (2007) 357–361. See also the posts by "Retread" at http://blogs.nature.com/thescepticalchymist/2008/04/chemiotics_how_many_proteins_c.html.

we are faced with the problem of how the 'few' extant proteins were produced and/or selected during prebiotic molecular evolution ... Even knowing a useful method to produce proteins under prebiotic conditions the problem would not be solved. In fact, for example, synthesizing a random 50 mer chain using all 20 different amino acids it is theoretically possible to produce about 10^{65} different sequences and the probability to sample two identical chains is approximately equal to zero."

Note that Chiarabelli and Lucrezia speak of proteins being "produced and/or selected during *prebiotic* molecular evolution". This means before the advent of self-reproducing organisms having not only the ability to replicate but also having a form of metabolism. When and how the first proteins came into being is a critical issue for the origin of life and will be further discussed below.

The argument that the number of proteins found in nature could not have arisen by Darwinian selection of random variations brings to mind the Levinthal paradox,[11] which has to do with protein folding times. Levinthal's paradox results from assuming an unbiased random search, and can be resolved by introducing a small energy cost for locally incorrect bond configurations.[12] In this way, the search is transformed into a biased search, which can dramatically reduce the number of configurations that need be sampled to arrive at a useful one.

From an evolutionary perspective, perhaps the most important proteins are the catalysts known as enzymes. What the probabilistic argument above tells us is that, if one assumes that the synthesis of each of the 20^{200} protein possibilities is equally probable, some of the earliest proteins capable of serving as enzymes could not have arisen from random selection from the set of all possible proteins. That is, they could not result from random variations in the sequence of amino acids followed by the natural selection of primitive organisms

---

[11]Levinthal, C., "Mossbauer spectroscopy in biological systems", in *Proceedings of a Meeting held at Allerton House, Monticello, IL*, eds. Debrunner, P., Tsibris, J. C. M. & Munck, E. (University of Illinois Press, Urbana, 1969), pp. 22–24.
[12]Zwanzig, R., Szabo, A. and Bagchi, B., "Levinthal's paradox", *Proc. Natl. Acad. Sci. USA* **89** (1992), 20–22.

utilizing the resulting proteins. The resolution of this conundrum must lie with the origin of early self-reproducing systems. Like the Levinthal paradox, the evolution of a set of biologically useful proteins could not have depended on an unbiased selection from all possible proteins. The argument also tells us that during the early development of life, the set of possible proteins had to be sampled in a massively parallel manner.

The way cells currently produce proteins is the result of some three billion years of evolution. Even a cursory examination of the process shows that it is far too complex to have been used by early life: To begin with, amino acids go through the process of "activation", in which an amino-acid-specific aminoacyl-tRNA synthetase combines a given amino acid with adenosine monophosphate (AMP) — one of the nucleotides that make up RNA — and attaches the activated amino acid to one of some 35 different types of tRNA. These in turn are brought to the ribosome via the guanosine triphosphate (GTP)/guanosine diphosphate (GDP) cycle where mRNA (after the introns are spliced out) instructs the ribosome to produce a specific polypeptide chain. While a majority of proteins produced in this way may fold on their own, an important number use auxiliary folding proteins. This entire process could only have sprung full blown into existence by the intervention of some *deus ex machina*!

The principal difficulty with the evolutionary approach to understanding the problem with protein development is that while the synthesis of amino acids in the reducing atmosphere of the early earth was demonstrated in the laboratory as early as 1953 by Miller,[13,14] it is energetically unfavorable for amino acids to combine

---

[13]Miller, S.L. "A production of amino acids under possible primitive earth conditions", *Science* **117** (1953), 528–529. A recent reanalysis of the residues of one of the original Miller experiments using modern techniques [Johnson, A.P. *et al.*, "The Miller volcanic spark discharge experiment", *Science* **322** (2008), 404.] showed the presence of 22 amino acids and five amines rather than the five amino acids originally found by Miller.

[14]Miller, S.L. and Orgel, L.E., *The Origins of Life on the Earth* (Prentis-Hall, Inc., Englewood Cliffs, N.J., 1974).

into polypeptides without the help of a catalyst. In modern cells, peptide bond formation is mediated by the energy in the amino acid-tRNA bond.

As an aside, Miller used electric discharges, to simulate lightning on the early Earth, to induce chemical reactions. Recently, Hess *et al.*[15] proposed lightning strikes on the prebiotic Earth to generate reactive phosphorus, which is essential for the synthesis of organic phosphate molecules. In their words, lightning strikes provide "a mechanism independent of meteorite flux for continually generating prebiotic reactive phosphorus on Earth-like planets, potentially facilitating the emergence of terrestrial life indefinitely".

There is some recent work that could shed some light on the issue of protein development. In studying the origin and evolution of the ribosome Harish and Caetano-Anollés[16] found that modern protein synthesis may have evolved from preexisting functions of primordial molecules. They found that universally conserved, functionally important components at the interface of the ribosomal small subunit and the large subunit are primordial.

It was the discoveries beginning in the early 1980s — that RNA could not only catalyze RNA replication but also direct peptide synthesis — that led to the idea of an "RNA World"[17] where life forms based on RNA existed before the ability to synthesize proteins from information encoded into DNA evolved. It was the recognition that the information contained in many eukaryotic genes was not contiguous and that introns had to be removed from the mRNA derived from these genes before the mRNA could be used for protein synthesis by the ribosome that led to these discoveries. Some RNAs were even found to be capable of self-splicing. Since the early work in

[15]Hesss, B.L., Piazolo, S., and Harvey, J., "Lighting strikes as a major facilitator of prebiotic phosphorus reduction on early Earth", *Nat. Commun.* **12** (2021), 1535.

[16]Harish, A. and Caetano-Anollés, G., "Ribosomal history reveals origins of modern protein synthesis", *PloS ONE* **7**(3) (2012), e32776. Doi:10.1371/journal.pone.0032776.

[17]Orgel, L.E., "Prebiotic chemistry and the origin of the RNA world", *Crit. Rev. Biochem. Mol. Biol.* **39** (2004), 99–123.

this area, catalytic RNAs (known as ribozymes) have been identified that form amide bonds (known as peptide bonds in a biochemical context) between RNA and an amino acid or between two amino acids.[18]

Both RNA-catalyzed aminoacyl-RNA synthesis and RNA-catalyzed amino acid activation have now been shown to be possible. Aminoacyl-tRNA synthetases catalyze two essential reactions: the activation of the carbonyl groups of amino acids by forming aminoacyl-adenylates designated as *aa-AMP* and the transfer of the aminoacyl group to a specific RNA. The first can be written as

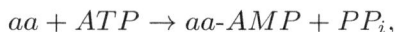

$$aa + ATP \rightarrow aa\text{-}AMP + PP_i,$$

where *ATP* and *AMP* stand for adenosine triphosphate and adenosine monophosphate respectively. The anion $P_2O_7^{4-}$ is abbreviated as PP_i and is formed by the hydrolysis of ATP into AMP. This reaction is followed by the transfer of the aminoacyl group to RNA

$$aa\text{-}AMP + RNA \rightarrow aa\text{-}RNA + AMP.$$

RNA-catalyzed aminoacyl-RNA synthesis was first achieved by Illangasekare, *et al.* in 1995[19] and amino acid activation by Kumar and Yarus in 2001.[20] The structural basis for specific tRNA aminoacylation by a small ribozyme has been given by Hong *et al.*[21]

There is also at least one plausible prebiotic polymerization reaction that can produce peptides. It has been shown by Leman, Orgel, and Reza Ghadiri in 2004[22] that the volcanic gas carbonyl

---

[18]Zhang, B. and Cech, T.R., "Peptidyl-transferase ribozymes: *trans* reactions, structural characterization and ribosomal RNA-like features", *Chem. Biol.* **5** (1998), 539–553.

[19]Illangasekare, M., Sanchez, G., Nickles, T. and Yarus, M., Aminoacyl-RNA synthesis catalyzed by an RNA", *Science* **267** (1995), 643–647.

[20]Kumar, R.K. and Yarus, M., "RNA-catalyzed amino acid activation", *Biochemistry* **40** (2001), 6998–7004.

[21]Hong, X., Murakami, H., Suga, H., and Ferré-D'Amaré , A.R., "Structural basis of specific tRNA amonoacylation by a small *in vitro* selected ribozyme", *Nature* **457** (2008), 358–361.

[22]Leman, L., Orgel, L. and Reza Ghadiri, M., "Carbonyl sulfide–mediated prebiotic formation of peptides", *Science* **306** (2004), 283–286.

sulfide is capable of polymerizing amino acids to form peptides under reasonable conditions. Whether this reaction could be the basis for a complementary or alternative scenario to that of the RNA world has yet to be determined.

This brings us back to the issue of whether life first began in the form of creatures capable of rudimentary metabolism coupled with replication or, as put by Dyson in his delightful little book *Origins of Life*,[23] whether "life began twice, with two separate kinds of creatures, one capable of metabolism without exact replication, the other kind capable of replication without metabolism". There is also Wong's coevolution theory of the genetic code[24] based on the postulate that prebiotic synthesis was an inadequate source of all twenty protein amino acids, and consequently some of them had to be derived from the coevolving pathways of amino acid biosynthesis.

Dyson introduced a "Toy Model" of molecular evolution along the lines of the Oparin picture of the origin of life[25] where proto-cells came first, followed by enzymes, and subsequently by the use of nucleic acids to store biological information. The model does not allow Darwinian selection. Its purpose was to demonstrate that a population of molecules within a proto-cell could, by random drift, achieve an organized state where active biochemical cycles might exist. Wong's theory, on the other hand, allows for a prebiotic evolution of peptide sequences as well as amino acid biosynthesis, and therefore at least partly makes use of Darwinian selection.

Without Darwinian selection, one cannot preferentially increase the number of proto-cells containing biologically useful molecules formed by random drift. Darwinian evolution can be introduced into Dyson's Toy Model by allowing the proto-cells to grow and reproduce by fission and dissolve in the absence of adequate nutrients. The precise extent and nature of prebiotic molecular evolution in the development of life is unknown, but a resolution of this question is

---

[23]Dyson, F., *Origins of Life* (Cambridge University Press, Cambridge, 1985).

[24]Wong, J., "Coevolution theory of the genetic code at age thirty", *BioEssays* **27** (2005), 416–425.

[25]Oparin, A.I., *Life: Its Nature, Origin, and Development* (Academic Press, Inc., New York, 1966).

not central to the issue at hand; being able to form proteins via the carbonyl sulfide or a similar route on the prebiotic Earth, or in an RNA World for that matter, does not in and of itself resolve the probabilistic argument given above.

Certain assumptions that may not at first be apparent are built into that argument. The first is that the evolution of life depends on selecting for the specific set of proteins that appears in life forms today. It seems very unlikely that the origin and evolution of life could be so narrowly constrained. The second is that each of the 20^{200} possible proteins is functionally unique. This is also unlikely to be the case. It is far more probable that many of these possible 20^{200} proteins can serve the same biological function. In the case of enzymes, it may have been possible for many structurally unrelated enzymes to catalyze a given reaction, albeit with possibly different reaction rates. This is certainly the case today. Such enzymes are known as "analogous" enzymes and represent independent paths in enzyme evolution.[26,27] This contrasts with homologous enzymes, which derive from a common ancestor. It is also probable that many of the enzymes of the RNA World were far less efficient than their modern cognates.

Another assumption implicit in the argument that the protein space to be sampled is $\sim 20^{200}$ is that an enzyme can catalyze only a single reaction. The active site of an enzyme is generally only a very small part of the protein and is usually formed by different polypeptide chains. Even the configurations of enzymes often change in the presence of the substrate molecule with which it interacts. It is quite possible that many of the 20^{200} proteins, especially the larger ones, could have several active sites catalyzing different reactions. In fact, many enzymes are known to have a number of catalytic sites.[28]

---

[26]Galperin, M.Y., Walker, D.R. and Koonin, E.V., "Analogous enzymes: Independent inventions in enzyme evolution", *Genome Res.* **8** (1998), 779–790.

[27]Gherardini, P.F., Wass, M.N., Helmer-Citterich, M. and Sternberg, M.J.E., "Convergent evolution of enzyme active sites is not a rare phenomenon", *J. Mol. Biol.* **372** (2007), 817–845.

[28]Llewellyn, N.M. and Spencer, J.B., "Enzymes line up for assembly", *Nature* **448** (2007), 755–756.

Taking account of these built in assumptions could dramatically reduce the set of possible proteins that need be sampled to produce an enzyme with a given activity.

There is another even more compelling argument that the set of proteins that was sampled during the early evolutionary process was far smaller than 20^{200}. Nineteen of the 20 amino acid residues in the secondary structures of proteins show a relatively strong tendency to form either α-helices, β-sheets, or turns.[29] That is, amino acids have different conformational propensities for forming these structures. Thus, from the secondary structural point of view, amino acids fall into three "fuzzy" equivalence classes. "Fuzzy" because, strictly speaking, an equivalence relation partitions a set into subsets where each member of the original set belongs to only one member of the partition. Here, amino acid residues only have a "tendency" to form α-helices, β-sheets, or turns.

In addition, it has been shown[30] that the secondary structure of proteins depends not only on the composition but also the sequence of amino acids in the protein. Some sequence patterns produce α-helical structures while others produce β-strands.

Most naturally occurring proteins are composed of between 50 and 2000 amino acids. If the size of early proteins was toward the lower end of this range, and if the tertiary structures upon which the protein's properties depend were strongly influenced by this division into three fuzzy amino acid equivalence classes, the set of proteins that would need to be sampled could be reduced. This reduction, in conjunction with the rules that specify the secondary structure of proteins, would dramatically reduce the sequence space that needs to be sampled. While the remaining number of proteins may still be large, it is one that could well be sampled in a reasonable time if the set of proteins is sampled in a parallel manner as discussed below.

---

[29] Berg, J.M., Tymoczko, J.L. and Stryer, L., *Biochemistry* (W. H. Freeman and Co., New York, 2002), 5th edition, §9.2.

[30] Wei, Y., Kim, S., Fela, D., Baum, J. and Hecht, M. H., "Solution structure of a *de novo* protein from a designed combinatorial library", *Proc. Nat. Acad. Sci.* **100** (2003), 13270–13273.

Another consideration that could reduce the set of possible proteins is the recognition that more than one-third of all enzymes contain either bound metal ions or require the addition of such ions for catalytic activity. Since 1932, when carbonic anhydrase — a biologically important zinc containing enzyme — was discovered, hundreds of enzymes have been found to contain zinc, but only in the +2 state.

Zinc atoms in such enzymes are essentially always bound to four or more ligands. Carbonic anhydrase, which catalyzes carbon dioxide hydration, is particularly important in biological systems. The way this enzyme probably works is that the zinc creates a hydroxide ion from a water molecule by facilitating the release of a proton; the carbon dioxide substrate then binds to the enzyme's active site where it is positioned to react with the hydroxide ion; the hydroxide ion then converts the carbon dioxide molecule into a bicarbonate ion; and finally, the catalytic site is restored by the release of the bicarbonate ion and the binding of another water molecule.

A synthetic analog of this carbonic anhydrase mechanism, where a simple synthetic ligand binds zinc through four nitrogen atoms — rather than the three histidine nitrogen atoms in the enzyme — accelerates the hydration of carbon dioxide by more than a factor of 100 at a pH of 9.2.[31]

Zinc is not the only metal ion of interest. It has also been proposed[32] that specific short peptides three to eight amino acids long bound to one or more positively charged metal ions such as Mg^{2+} could have served as catalysts during the period of very early life.

What this suggests is that early proto-enzymes may have contained metal complexes that were subsequently incorporated into evolving protein enzymes. If true, this too could have diminished the number of possible proteins that need be sampled — provided large

[31] Berg, J.M., Tymoczko, J.L. and Stryer, L., *Biochemistry* (W. H. Freeman and Co., New York, 2002), 5th edition, §9.2.

[32] van der Gulik, P., Massar, S., Gilis, D., Buhrman, H. and Rooman, M., "The first peptides: The evolutionary transition between prebiotic amino acids and early proteins", *J. Theor. Biol.* **261** (2009), 531–539.

subsets of the possible proteins were able to incorporate a given metal complex.

Thus, what the probability argument is really saying is that the possibility of life arising for a second time in exactly the same way it did is negligble. This may well be true, but many alternative possibilities may have existed for life to begin in the form of self-replicating systems. It is even possible that more than one type of self-replicating organism appeared and all were subject to variation, selection, and in the end, possible convergence.

The idea of an RNA World has its own difficulties. Two of the most important are obtaining, under pre-biotic conditions, nucleotides in sufficient quantity, and the lack of a known mechanism for replication of RNA molecules without the presence of a replicase. These challenges have not yet been fully resolved. However, once self-replicating RNA molecules capable of catalyzing polypeptide chains exist, it would be expected that different strands of RNA would produce different polypeptides. Collectively, these effectively sample the set of proteins in a massively parallel manner. Once the minimal necessary set of ribozymes were formed, even if their reaction rates were much lower than their modern cognates, RNA would serve to carry both "genetic" information and serve as a catalyst for the reactions needed for the first primitive self-replicating systems.

These primitive RNA World proto-cells would be subject to the usual Darwinian variation and selection process. There is, however, a strong incentive for an evolutionary transition from an RNA to a DNA world.[33] This is due to the greater genetic stability of the double-stranded helical structure of DNA compared to single-stranded RNA, and the fact that deoxyribose is chemically less reactive than ribose.

In all modern organisms, the components of DNA are synthesized from those of RNA by ribonucleotide reductases. These convert the base and phosphate groups linked to a *ribose* sugar that form ribonu-cleotides to deoxyribonucleotides composed of a base and phosphates

[33]Lazcano, A., Guerrero, R., Margulis, L. and Oró, J., "The evolutionary transition from RNA to DNA in early cells", *J. Mol. Evol.* **27** (1988), 283–290.

linked to *deoxyribose* sugar. The fact that ribonucleotide reductases take a variety of species-dependent forms increases the probability that a primordial form of these reductases could have formed in the RNA World. Another possibility is that ribozymes were replaced by enzymes formed of amino acids before the evolutionary transition to storing information in DNA.

The transition to a DNA World is complicated by the fact that some mechanism for reading the information from the DNA would have had to evolve along with the transition to information storage in DNA.

One approach to resolving this problem might be along the following lines: The two strands that comprise DNA separate on heating, and reform on cooling. If RNA is present during the cooling period, DNA–RNA hybrids will form where sections of the DNA and the RNA have complementary bases. Under natural selection, primitive self-reproducing proto-cells of the RNA World would have a preponderance of the type of RNA needed to support their metabolism and replication. If some of these RNAs were converted to DNAs by early forms of ribonucleotide reductases, a replication of RNAs using these DNAs as a template by temperature cycling may serve in place of the modern enzyme RNA polymerase. Once such DNA-based self-replicating systems existed, natural selection would rapidly end the RNA world.

It is interesting, and relevant to this discussion, that the Central Dogma has itself been questioned. Francis Crick, who coined the phrase in 1957, apparently views RNA editing as the most significant exception.[34] Such editing involves the modification of RNA sequences after transcription and is a common occurrence in eukaryotic cells. Although editing implies that the sequence of amino acids in the resulting protein is not entirely encoded in nucleic acids, there is still no known mechanism where the information for an amino acid sequence could flow from protein to nucleic acid.

---

[34]Thieffry, D. and Sarkar, S., "Forty years under the central dogma", *Trends Biochem. Sci.* **23** (1998), 312–316.

RNA editing is common in mRNA transcribed from mitochondrial DNA. It was soon discovered, however, that the sequence information needed for RNA editing was supplied by small RNAs transcribed from a second component of the mitochondrial DNA. If this turns out always to be the case, RNA editing would involve the transfer of genetic information from one RNA to another.[35] This would resolve the editing challenge to the Central Dogma. Other challenges to the Central Dogma, such as the discovery of reverse transcriptase, the replication of prions, and epigenetic modulation of DNA, have been discussed elsewhere — for example, by Morange.[36]

To summarize: If proteins on the average are composed of 200 amino acids, one possible way out of the dilemma of selecting the 20,000 or so proteins coded for in our genes (or the larger number found in the natural proteome) out of the 20^{200} possible is to reduce the set of possible proteins by recognizing that: it may have been possible for organisms that first populated the post-RNA World to use many different subsets of the 20^{200} possible proteins; when these proteins are enzymes, many more than one of them may catalyze any given reaction, although possibly with differing reaction rates. In addition, many proteins may have more than one active site and thus serve to catalyze more than one reaction. Also, of the 20^{200} possible, many proteins may be able to serve the same biological function. The set of proteins that need be sampled may also have been reduced if many of the early protein-based enzymes were able to incorporate the metal complexes that may have constituted earlier proto-enzymes. And, finally, if the tertiary structure of early proteins was strongly influenced by the division of amino acids into three fuzzy equivalence classes, the number of proteins that need be sampled could be closer to $\sim 3^{50}$ rather than 20^{200}. The basic point is that the evolution of proteins could not, and need not have involved an unbiased random search of all possible proteins.

---

[35]Seiwert, S.D., "RNA editing hints of a remarkable diversity in gene expression pathways", *Science* **274** (1996), 1636–1637.
[36]Morange, M., "Fifty years of the Central Dogma", *J. Biosci.* **33**(2) (2008), 171–175.

Once the transition to a DNA world is accomplished, the set of possible proteins would continue to be sampled by variation and selection of early DNA-based cells. These, as many single-celled organisms do today, would presumably share genes through horizontal gene transfer. Because of this mode of information exchange, the evolutionary process of variation and selection becomes a massively parallel rather than a serial process. This is what allows colonies of microorganisms to rapidly adapt to changing environmental conditions today.

The following three chapters will explain why the rise of life is not a stochastic process.

Chapter 3

The Role of Prebiotic Molecules and Protocells in the Origin of Life

The issue of abiogenesis has an enormous literature that convincingly shows that the origin of life was not a stochastic process. To begin with, our galaxy has enormous clouds of organic molecules the most abundant of which are carbon monoxide and formaldehyde. There is also considerable nitrogen and water. In addition, amino acids, the building blocks of proteins, have been found in meteorites, of which the most famous is the Murchison carbonaceous chondrite meteorite. Eight of the approximately 86 amino acids found are also found in known terrestrial proteins.

Ribonucleic acid (RNA) can be synthesized under abiotic conditions and, in the early 1980s, it was discovered that not only could RNA catalyze RNA replication but also catalyze the synthesis of a peptide bond that led to the idea of an RNA world. Catalytic abiotic RNAs, known as ribozymes, have been experimentally "evolved" in the laboratory and are capable of peptide bond formation, nucleotide synthesis, and the cleavage and joining of RNA. Self-replicating peptides have also been found, implying that abiotic formation of short peptides, some of which are chiral selective during self-replication, could have been available on the prebiotic Earth.

Once self-replicating RNA molecules capable of catalyzing polypeptide chains existed, different strands of RNA would produce different polypeptides. Collectively, these could effectively sample many, if not all, possible proteins in a massively parallel manner. When the minimal necessary set of ribozymes were formed multiple short RNAs could serve to carry both "genetic" information and serve as catalysts for the reactions needed for the first primitive self-replicating systems.

Life not only requires the existence of replicating molecules but also a cellular membrane that would separate the contents of a protocell from the surrounding environment. It would also mediate the selective exchange of solutes. Early single-layer membranes (micelles) and bilayer membranes (vesicles) would use molecules that were able to self-assemble under appropriate conditions (Figure 1).

Phospholipids as well as glycolipids tend to form vesicles rather than micelles because the two chains making up their hydrophobic tail are too large to fit into the interior of a micelle, which is limited in size. Vesicles are not so limited and can have macroscopic sizes such as a millimeter. A short introduction to prebiotic molecules and the formation of protocells has been given by Marsh[1] and a much more extensive review by Meierhenrich *et al.*[2]

A review of physical and chemical heterogeneity and membrane self-assembly has been given by Szostak[3] who shows that "membrane self-assembly under simple and natural conditions gives rise to heterogeneous mixtures of large multi-lamellar vesicles, which are predisposed to a robust pathway of growth and division that simpler and more homogeneous small unilamellar vesicles cannot undergo."

The example of the self-assembly of amphiphilic molecules is not unusual. Self-assembly of what are called supramolecular complexes such as viral capsids often occur in nature. Viral capsids are one-molecule-thick shells of protein usually containing a single copy of a nucleic acid genome. Capsids can have a complicated structure such as in the case of a bacteriophage or more simply a polyhedron. Retroviruses also have a limited number of enzymes. The physics of

---

[1]Marsh, G.E., *The Immense Journey* (World Scientific Publishing Co., 2018), Chapter IV (pp. 75–85 of Chapter V are also relevant).

[2]Meierhenrich, U.J. *et al.*, "On the origin of primitive cells: From nutrient intake to elongation of encapsulated nucleotides", *Angew. Chem. Int. Ed.* **49** (2010), 3738–3750.

[3]Szostak, J.W., "An optimal degree of physical and chemical heterogeneity for the origin of life?", *Phil. Trans. R. Soc. B* **366**, (2011), 2894–2901.

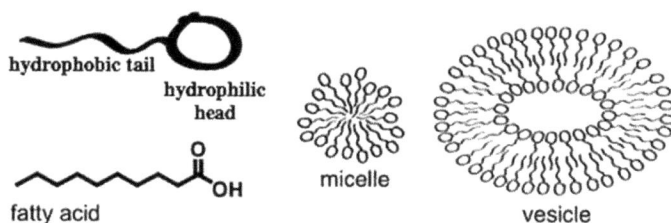

Figure 1. Self-assembly of amphiphilic molecules (here a fatty acid) into a micelle and a vesicle. The hydrocarbon chain making up the hydrophobic tail is indicated by the symbol ∿∿ representing the single covalent bonds between the carbon atoms that make up the backbone of the tail. In general, amphiphilic molecules could have multiple hydrocarbon tails. The one shown in the figure is a fatty acid with only one tail. (G.E. Marsh, *The Immense Journey.*)

self-assembly and its history has been given by Bruinsma *et al.*[4] Viruses are of interest here because one possibility proposed for their origin is that the ancestors of modern viruses were free-living noncellular predecessors of cellular organisms. From this point of view, viruses were a nearly-living stage in the origin of life.[5] What is clear is that viruses and their cellular hosts are evolutionarily intertwined due to horizontal gene transfer.

Returning to the subject of cellular membranes, one remaining question is how could large molecules be captured by prebiotic vesicles? The answer has been given by Deamer[6] and confirmed by experiment. He showed that the permeability barrier could be broken down by wet and dry cycles applied to phospholipid vesicles mixed with large molecules such as proteins or nucleic acids.

This means that life could have begun with replicating protocells that would self-assemble from molecules found on the primitive Earth and would not require an external source of energy. Such protocells could be the basis for the origin of the first primitive prokaryotes.

---

[4]Bruinsma, R.F., Wuite, G.J.L. and Roos, W.H., "Physics of viral dynamics", *Nat. Rev. Phys.* **3** (2021), 79–91.

[5]Moelling, K. and Broecker, F., "Viruses and evolution-viruses first? A personal perspective", *Front. Microbiol.* **10** (2019).

[6]Deamer, D., *First Life* (University of California Press, Berkely 2011), Chapter 8.

Along a path taken by Oparin[7] and Haldane[8] who claimed that a protocell was a proliferating droplet known as a coacervate, Matsuo and Kurihara[9] constructed a self-reproducing droplet using liquid–liquid phase separation (LLPS).

They credit the inspiration for the procedure they used to de Duve's "thioester world" hypothesis,[10] which argues that prebiotic peptides might have been generated from amino acid thioesters under mild, aqueous conditions. In their words, they

> "were able to simultaneously form droplets via LLPS and generate peptides by using a designed and synthesised thioesterified cysteine derivative as a monomer precursor for the spontaneous ligomerisation of an amino acid thioester under mild, aqueous conditions. A continuous supply of a monomer precursor that kept the LLPS-formed droplets in a non-equilibrium state enabled the LLPS-formed droplets to undergo a steady growth-division cycle that maintained droplet size while increasing the number of droplets. We also showed that LLPS-formed droplets were able to resist dissolution by lipids and to maintain themselves when nucleic acids and lipids were both present in them if the concentrated nucleic acids were localised at the inner boundary of the LLPS-formed droplet with the assistance of generated peptides. Overall, we were able to demonstrate how a proliferating droplet protocell could be formed by the oligomerisation of amino acid thioesters and functionalised by oligonucleotides. Such a protocell could have served as a link between "chemistry" and "biology" during the origins of life. This study may serve to explain the emergence of the first living organisms on primordial Earth."

There have been many experimental approaches designed to ultimately create life in the laboratory. Functioning virus genomes

---

[7]Oparin, A.I., *The Origin of Life*, Bernal, J.D., ed. (Weidenfeld and Nicholson, 1967), pp. 199–234.

[8]Haldane, J.B.S., "Origin of life", *Rationalist Annu.* **148** (1929), 3–10.

[9]Matsuo, M. and Kurihara, K., "Proliferating coacervate droplets as the missing link between chemistry and biology in the origins of life", *Nat. Commun.* **12** (2021), Article number 5487.

[10]de Duve, C. *Blueprint for a Cell: The Nature and Origin of Life* (Carolina Biological Supply Co. Neil Patterson, 1991).

have already been created some time ago from scratch,[11] but the creation of a minimal functioning cell, including the cytoplasm, is still elusive. At this time, it may be the complexity of the cytoplasm that presents the most difficulty. Here is a description of the cytoplasm given by Luby-Phelps[12]:

> "The cytoplasmic compartment is inhomogeneous at nearly every length scale. In addition to randomly distributed local inhomogeneity driven stochastically by crowding and phase separation, nonrandom localization of intracellular vesicles, organelles, and supramolecular assemblies is a hallmark of eukaryotic cells. It is becoming clear that in prokaryotes, as well as in eukaryotes, individual protein and RNA molecules may also be nonrandomly localized within the cytoplasmic compartment. An extensive literature suggests that the concentrations of even small signaling molecules such as cAMP and Ca^{2+} may be locally regulated. It is important to remember that reported values for the physical properties of cytoplasm are spatially and temporally averaged and thus may not well describe the conditions in any particular subvolume of the cell."

It would be even more difficult to create a eukaryotic cell than a minimal functioning cell. Yet, in 2018 the first functional synthetic eukaryotic genome, known as Sc2.0, was created. It is a modified *Saccharomyces cerevisiae* (common baker's yeast) genome that was reduced in size by about 8%, and where a series of edits that involved deletions, insertions, and base substitutions were performed. There are 16 native *S. cerevisiae* chromosomes. Over seven of these synthetic chromosomes were successfully synthesized and combined in a single yeast cell to produce a strain with more than 50% synthetic DNA that survives and replicates.[13,14] As an aside, yeasts are classified as members of the fungus kingdom.

---

[11]Gibsom, D.G. *et al.*, "Creation of a bacterial cell controlled by a chemically synthesized genome", *Science* **329** (2010), 52–56.
[12]Luby-Phelps, K., "The physical chemistry of cytoplasm and its influence on cell function: An update", *Mol. Biol. Cell* **24** (2013), 2593–2596.
[13]Truong, D.M. and Boeke, J.D., "Resetting the yeast epigenome with human nucleosomes", *Cell* **171** (2017), 1508–1519.
[14]Richardson, S.M. *et al.*, "Design of a synthetic genome", *Science* **355** (2017), 1040–1044.

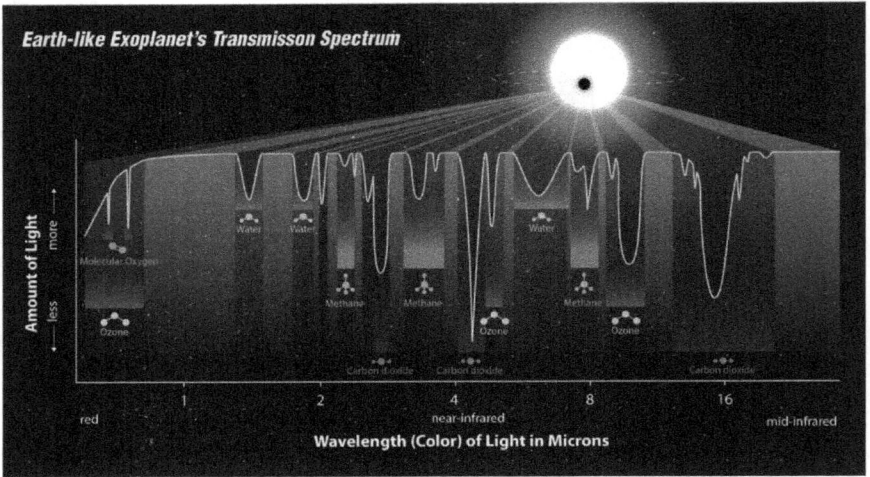

Figure 2. Simulated transmission spectrum through an Earth-like planet's atmosphere. (NASA, STSci.)

Even if a fully functioning minimal cell was created in the laboratory that would still not definitively prove that life on Earth-like planets is inevitable. Many would only accept clear observational evidence of life on Earth-like planets to be definitive. This would most likely come from looking at the transmission spectrum through the atmosphere of an Earth-like planet. NASA's James Webb Telescope is currently the only instrument with any hope of successfully making such an observation. Figure 2 shows a 2021 simulation of such a transmission spectrum.

The dips in the simulated spectrum show the wavelengths of the starlight where the light is absorbed. These features would be extremely faint and at the limits of the James Webb Space Telescope's sensitivity.

Even if most might accept abiogenesis on Earth-like planets without definitive scientific proof, the debate has recently shifted to eukaryogenesis, which is often argued to be an improbable evolutionary transition unique to our Earth. On Earth, only eukaryotes have evolved into complex life.

Chapter 4

The Prokaryote–Eukaryote Divide

This chapter will show that the development of complex life on Earth-like planets cannot be ruled out by what is known as the prokaryote–eukaryote divide. The argument leading to the prokaryote–eukaryote divide is based on the idea that prokaryotes cannot engulf other cells without phagocytosis. Most importantly, as will be seen below, the evolutionary transition from prokaryotes to eukaryotes is not based on a stochastic process and consequently one cannot use statistics to determine the probability of the evolution of eukaryotes.

Prokaryotes are single-celled organisms that lack a nucleus or any other membrane-bound organelle. In contrast, eukaryotic cells have a membrane-bound nucleus as well as other membrane-bound organelles that have specialized functions. The prokaryote–eukaryote divide dates back to the 1960s. The purpose it served and what it implied for classification and phylogeny has been covered in a very interesting history by Sapp.[1]

Endosymbiosis,[2] where one cell is engulfed by another and remains there as a useful symbiont, was argued to be the origin of eukaryotic cells by Lynn Margulis in the late 1960s and she published a book on the subject in 1998.[3] (See Appendix D for a more in-depth discussion of symbiosis.) The important symbiont for animal

---

[1]Sapp, J., "The prokaryote-eukaryote dichotomy: Meanings and mythology", *Microbiol. Mol. Biol. Rev.* **69** (2005), 292–305.

[2]Martin, W.F., Garg, S. and Zimorski, V, "Endosymbiotic theories for eukaryote origin", *Phil. Trans. R. Soc. B* **370** (2014), 330.

[3]Margulis, L., *Symbiotic Planet* (Sciencewriters, Mass., 1998).

cells was the endosymbiosis of the cell that led to the mitochondria. Mitochondria share a common ancestor with the alphaproteobacteria, a diverse class of organisms that include the most abundant of marine cellular organisms including the protomitochondrion, the extinct bacterium that evolved into the mitochondria. Only Eukaryotes have mitochondria, giving such cells orders of magnitude more energy per gene (a common measure), allowing the development of complex life. Eukaryogenesis was a major change in the development of life in that mitochondria and their host cells could only replicate together as a whole so that eukaryotes are subject to Darwinian variation and selection extended by the discoveries of Gregor Mendel to form the *Modern Synthesis*. As an aside, an additional extension of the *Modern Synthesis*, by including transgenerational epigenetic inheritance, is still being formulated and is often called *The Extended Synthesis*.

It is often said that the origin of the eukaryotic cell some two billion years ago was a unique or "singular" evolutionary event as argued by Lane[4] in his comprehensive article "Energetics and genetics across the prokaryote-eukaryote divide". This idea has been explained in great detail in Lane's extraordinary book *The Vital Question*.[5] He states in no uncertain terms that the origin of eukaryotes was due to "a single chimeric event between an archaeal host cell and a bacterial endosymbiont", which evolved into mitochondria. Bacteria and archaea are both prokaryotes. Endosymbiosis is rare between prokaryotes because it is believed that prokaryotes cannot engulf other cells without phagocytosis. The use of the phrase "single chimeric event" has led to some confusion in the literature, with a clarification given by Booth and Doolitle[6] that what Lane meant by a single chimeric event is that there was only one last eukaryotic common ancestor (LECA).

Genomic analyses show that the type of cell that absorbed the bacterial endosymbiont was most likely an ancient archaeon.

---

[4]Lane, N., Lane, *Biology Direct* **6** (2011), 35–66.

[5]Lane, N, *The Vital Question* (W.W. Norton & Co. Inc., New York, 2015).

[6]Booth, A. and Ford Doolittle, W., "Eukaryogenesis, how special really?", *PNAS* **112** (2015), 10278–10285.

Some have argued against the idea of an endosymbiosis between two prokaryotes because prokaryotes are not capable of phagocytosis. But there are known examples of prokaryotic cells hosted by other prokaryotes, so while it may be relatively rare it does happen.

Lane argues that the endosymbiosis of the prokaryote that evolved to become the mitochondrion gave eukaryotes a vast increase in the amount of adenosine triphosphate (ATP) available for cellular use, which enabled eukaryotes to become more complex; ATP is the molecule used to store energy by all extant life forms. The availability of this additional energy was then the fundamental reason for the rise of the eukaryotes.

It is interesting there are no surviving evolutionary intermediates between prokaryotes and eukaryotes, an evolution which took place over a time period of about 2.1 billion years or so. Lane explains this large gap in time by noting the fact that there are simple eukaryotes today without mitochondria. This means that the early eukaryotes that did gain mitochondria constituted a small unstable population that slowly accumulated new traits and became more stable as they evolved to become closer to the LECA. In the end these earlier eukaryotes died out leaving the large time gap between prokaryotes and eukaryotes.

Eukaryogenesis is widely believed to be a statistically improbable evolutionary event unique to our planet. If this were true, it would mean that complex life on Earth-like planets would be very unlikely.

There are two principal theories about the origin of mitochondria and chloroplasts in eukaryotes. The first is phagotrophic engulfment and the second is microbial symbiosis. The two differ about which came first mitochondria or phagocytosis.

There are two examples of energy producing components in eukaryotic cells that were derived by endosymbiosis. The first is the mitochondria and the second is the chloroplasts in plant cells, which came from cyanobacteria. In 1970, Stanier[7] placed the origin of chloroplasts before the origin of mitochondria maintaining that

---

[7]Stanier, Y., "Some aspects of the biology of cells and their possible evolutionary significance", *Symp.Soc. Gen. Microbiol.* **20**, (1970), pp. 1–38.

since mitochondria use oxygen and because eukaryote origin took place during the anaerobic period, there must have been first an adequate source of oxygen before mitochondria were able to develop.

By 2015, Booth and Doolittle, referencing work by von Dohlen *et al.*[8] and Husnik *et al.*[9] noted that "prokaryotes can acquire other prokaryotes as endosymbionts without first developing eukaryote-like phagotrophy."

Also, in 2015 a clear review of endosymbiosis and eukaryotic cell evolution was given by Archibald.[10] With regard to photosynthesis he states that:

> "Despite the availability of genome sequence data from diverse algal lineages, key 'when' and 'how' questions about the evolution of eukaryotic photosynthesis remain. What we do know is that the host for the cyanobacterial progenitor of the plastid was a mitochondrion-containing eukaryote, a single-celled heterotrophic organism capable of ingesting prey by phagocytosis."

In 2020, Mills[11] found that

> "genomic and cytological evidence has increasingly supported the view that the pre-mitochondrial host cell — a bona fide archaeon branching within the 'Asgard' archaea — was incapable of phagocytosis and used alternative mechanisms to incorporate the alphaproteobacterial ancestor of mitochondria. Indeed, the diversity and variability of proteins associated with phagosomes across the eukaryotic tree suggest that phagocytosis, as seen in a variety of extant eukaryotes, may have evolved independently several times within the eukaryotic crown-group."[12]

---

[8]von Dohlen, C.D. *et al.*, "Mealybug β-proteobacterial endosymbionts contain γ-proteobacterial symbionts", *Nature* **412** (2001), 433–436.

[9]Husnik, F. *et al.*, "Horizontal gene transfer from diverse bacteria to an insect genome enables a tripartite nested mealybug symbiosis", *Cell* **153** (2013), 1567–1578.

[10]Arcdhibald, J.M., "Endosymbiosis and eukaryotic cell evolution", *Curr. Biol.* **25** (2015), R911–R921.

[11]Mills, D.B., "The origin of phagocytosis in Earth history", *Interface Focus* **10** (2020).

[12]In phylogenetics, the crown group is a collection of species composed of the living representatives of the collection, the most recent common ancestor of the collection, and all descendants of the most recent common ancestor.

More recently, Bremer *et al.*[13] in 2022 found that mitochondria preceded the origin of phagocytosis so that "phagocytosis cannot have been the mechanism by which mitochondria were acquired."

What is important here is that the implication of the above discussion is that eukaryogenesis is almost certainly not an improbable evolutionary event unique to our planet. Nor was eukaryogenesis stochastic in nature. Thus, the development of complex life on Earth-like planets cannot be ruled out by what is known as the prokaryote–eukaryote divide.

---

[13]Bremer *et al.*, "Ancestral state reconstructions trace mitochondria but not phagocytosis to the last eukaryotic common ancestor", *Genome Biol. Evol.* **14**(6). https://doi.org/10.1093/gbe/evac079 (accessed on 1 June 2022).

Chapter 5

Evolution of Polymers Under Abiotic Conditions

The argument has been made that the combing of the monomers associated with life into the polymers required for life must be a stochastic process. Were this to be true, the probability of life arising on Earth would have been vanishingly small. While the understanding of the evolution of polymers from monomers found in the prebiotic environment is not complete, there is already enough understood that would rule out a stochastic basis for the evolution of polymers. This chapter will give a limited introduction to the modern work on the evolution of polymers under abiotic conditions.

Most macromolecules are produced via the dehydration synthesis process, sometimes called the condensation reaction, and much work has been done to find means on the early earth that could be responsible for this synthesis.

For un-ionized monomers, a dehydration synthesis reaction allows the hydrogen of one monomer to combine with the hydroxyl group of another monomer, releasing a molecule of water in the process. An example is amino acids in an aqueous environment where two hydrogens from the positively-charged end of one monomer are combined with the oxygen from the negatively-charged end of another monomer, again releasing water and joining the two monomers with a covalent bond. The general form of an amino acid is shown in Figure 1.

The dehydration synthesis reaction between two amino acids is shown in Figure 2.

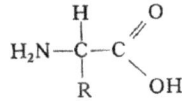

Figure 1. An amino acid. Each amino acid has a side chain, here designated as R.

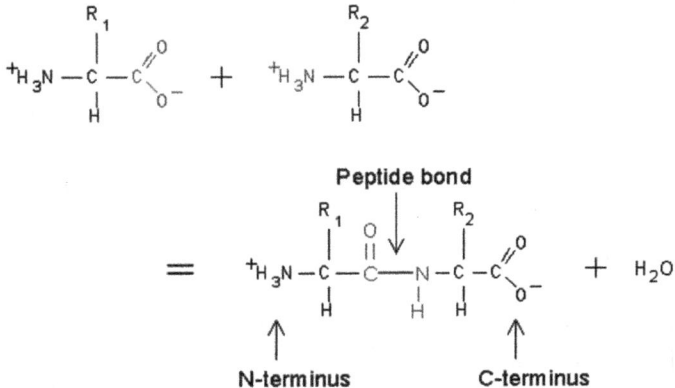

Figure 2. The dehydration synthesis reaction between two amino acids. All amino acids start with an amino group (NH_2) and end with a carboxyl group (COOH) and have differing other chemical groups attached indicated here by R_1 and R_2. This figure is often depicted by having the hydrogen ion on the left of the amino acid attached to the oxygen on the right so that the ends have no charge. Whether the ends are charged or not depends on the pH of the aqueous environment and in the figure above the pH is about 6, somewhat acidic. When more amino acids are attached to the dipeptide shown, the result is called a polypeptide. The ends (terminus) of the polypeptide are named for the nitrogen at the start and the carbon at the end.

See the discussion in Chapter 1 for how the dehydration synthesis reaction can become thermodynamically favorable by the process of wet-dry cycles on the early Earth.

Life today uses what is called the acetyl coenzyme A (acetyl CoA) pathway for the fixation of carbon (see Appendix B). It uses carbon dioxide and hydrogen to form small, reactive organic molecules, a process that releases energy. This energy is adequate to support the formation of nucleotides as well as that needed for polymerization of these molecules into the proteins RNA and DNA. An abiotic equivalent to acetyl CoA, which can supply metabolic energy, that is readily formed in alkaline hydrothermal vents is methyl thioacetate,[1] which can produce small organic molecules as well as nucleotides from carbon dioxide and hydrogen.[2] Methyl thioacetate can also react directly with phosphate to form acetyl phosphate, which can produce other organic molecules and polymerize these smaller molecules into long chains to form the proteins DNA and RNA. This is one of the reasons that many have come to believe that life originated in hydrothermal vents. While ribose is difficult to synthesize under abiotic conditions, it has been shown that ribose could have been formed by photochemical and thermal means in precometary ices in the protoplanetary disc.

S-Methyl thioacetate, also known as methyl thioacetic acid belong to the class of organic compounds known as thioesters. The accumulation of these compounds in various geological settings was examined by Chandru *et al.*[3] in a laboratory setting.

RNA can not only catalyze RNA replication but also catalyze the synthesis of peptide bonds. This led to the idea of an "RNA world" where life forms based on RNA existed before the ability to synthesize proteins from information encoded DNA evolved. That belief that RNA could be the first replicating molecule was due to the finding that RNA could not only hold genetic information but could also act as an "enzyme" so that RNA itself could catalyzes RNA replication. This strengthened the concept of an RNA world. Catalytic abiotic ribozymes have been experimentally evolved in the laboratory that are capable of peptide bond formation, nucleotide synthesis, and the cleavage and joining of RNA. The concept of the RNA world became the almost universally accepted hypothesis for the early stages of life and its evolution on Earth.

---

[1]Whicher, A., "simulated alkaline hydrothermal vent environments to investigate prebiotic metabolism at the origin of life". http://discovery.ucl.ac.uk/1515740/1/Whicher_Thesis.pdf.

[2]Martin, W. & Russell, M.J., "On the origin of biochemistry at an alkaline hydrothermal vent", *Phil. Trans. R. Soc. B* **362** (2007), 1887–1925.

[3]Chandru, K. *et al.*, "The abiotic chemistry of thiolated acetate derivatives and the origin of life", *Sci. Rep.* **6** (2016), 29883.

Soon, however, the argument was raised that the RNA polymer is too complex to have arisen solely from abiotic synthesis. But it is possible that the functions of ribozymes could have been assisted by peptides and other molecules present in the environment. This would have enhanced the stability, efficiency, and specificity of primitive ribosomes. Saad[4] calls this the "ribonucleopeptide world" at the origin of life.

The question of whether RNA is the product of evolution from a form of proto-RNA is the subject of a Cold Spring Harbor perspective by Engelhart and Hud.[5] They focus on candidates for proto-RNA that would have made prebiotic assembly relatively easy and where only few evolutionary changes would be required to produce RNA. They suggest the scenario where

> "a proto-RNA is more likely to have spontaneously formed than RNA, because a proto-RNA could have had more favorable chemical characteristics (e.g., greater availability of precursors and ease of assembly), but such a polymer was eventually replaced, through evolution, by RNA (potentially after several incremental changes), based on functional characteristics (e.g., nucleoside stability, versatility in forming catalytic structures). Thus, contemporary RNA may possess chemical traits that, although optimally suited for contemporary life, may have been ill-suited for the earliest biopolymers, with the converse being true for proto-RNA."

Chandru *et al.*,[6] point out that life may not have originated using the same monomeric components that it does today. There could have been many non-biological, what they call "xenobiological", monomers that were prebiotically available and capable of oligomerization and self-assembly. These would then presumably evolve into RNA.

---

[4]Saad, N.Y., "A ribonucleopeptide world at the origin of life", *J. Syst. Evolution* **56** (2018), 1–13.

[5]Engelhart, A.E. and Hud, N.V., "Primitive genetic polymers", *Cold Spring Harb Perspect Biol.* **2** (2010).

[6]Chjandru, K. *et al.*, "Prebiotic oligomerization and self-assembly of structurally diverse xenobiological monomers", *Sci. Rep.* **10** (2020), 17560.

Broecker[7] has also hypothesized the early existence of a protogenome consisting of multi-ribozyme RNA molecules that would evolve within liposomes formed in dry-wet cycling environments. These liposomes would be the first protocells. He notes that the transition from abiotic molecules to replicating entities subject to Darwinian evolution has two main hypotheses; the RNA world, or replicator first, and alternatively the one where metabolism arises first. In the RNA world, the ancestor of the genome arose spontaneously as a self-replicating oligomer (a molecule that consists of relatively few repeating units) or a polymer of RNA and that metabolism emerged as a consequence of the evolving RNA. Broecker suggests "that a protometabolism of catalytic RNAs (ribozymes) evolved first and gave rise to the first iteration of the genome, the protogenome."

There is also experimental evidence that lipid-encapsulated polymers (amphiphilic compounds are present in carbonaceous meteorites that could self-assemble into membranous vesicles) can be synthesized by cycles of hydration and dehydration. Cycling through wet, dry, and moist phases will subject polymers to a process of combinatorial selection that can result in structural and catalytic functions from initially random sequences. This would not take long since it would be done in a massively parallel manner. Note that the random sequences are not necessarily stochastic in nature, some amino acids may be incorporate more readily than others.[8]

It should also be noted that the oligomerization of amino acids under prebiotic conditions into polymers does not mean that the sequence of amino acids is predetermined, rather that the addition of an additional amino acid only depends on the current state of the growing polymer; i.e., it could be thought of as a Markov process (see Appendix A).

---

[7]Broecker, F., "Genome evolution from random ligation of RNAs of autocatalytic sets", *Int. J. Mol. Sci.* **22** (2021).

[8]Frenkel-Pinter *et al.*, "Selective incorporation of proteinaceous over nonproteinaceous cationic amino acids in model prebiotic oligomerization reactions", *PNAS* **116** (2019), 16338–16346.

Although somewhat dated at this time, for an in-depth discussion of the oligomerization of chemical and biological compounds, as well as a detailed treatment of the organic chemistry involved, see Kawamura's chapter titled "Oligomerization of Nucleic Acids and Peptides under the Primitive Earth Conditions".[9]

What this and the previous two chapters have shown is that the origin of life is not a stochastic process and that consequently the statistical approach to estimating the probability of life on Earth-like planets is invalid. As stated in the Preface, the appearance of life is founded on the ability of matter, governed by the principles of quantum mechanics, to form the molecules need for life to exist. On Earth-like planets, it is a natural process governed by the same laws as nonliving processes; it is the inevitable outcome of biochemical forces woven into the fabric of the universe.

---

[9]Lesieur, C. (ed.), *Oligomerization of Chemical and Biological Compounds* (InTech, 2014), Chapter 6. doi:10.5772/57075; http://dx.doi.org/10.5772/58222; http://intechopen.com/books/3847.

Chapter 6

Multicellular Life, Symmetry and The Morphology of Life on Earth-like Planets

After the development of photosynthesis, the early forms of which evolved around three billion years ago on Earth, the next most important step toward complex life was the development of multicellularity. There are five major types of complex multicellular organisms today — land plants, fungi, two types of algae, and animals — and it is now understood that these groups arose separately from different types of unicellular ancestors. Cells form two basic types of communal groupings, those where each cell is functionally equivalent and those where within such colonies the cells are differentiated by structure and function.

Fossils in ancient rock formations tell us that multicellularity dates back at least 3.5 billion years. These are filamentous fossils of prokaryotes likely to have been different types of oxygen-producing cyanobacteria. The oldest fossils of differentiated cell colonies date to about 2 billion years ago, although recent work suggests that multicellular eukaryotes (*Qingshania magnifica*) date back to 1.63 billion years ago.[1] The differentiation between the different types of oxygen-producing cyanobacteria is thought to be between nitrogen-fixing cells and photosynthetic cells. These fossils closely resemble contemporary cyanobacteria.

Each of the multicellular lineages, in order to maintain structural integrity, had to independently evolve a mechanism for cellular

---

[1]Miao, L. *et al.*, "1.63-billion-year-old multicellular eukaryotes from the Chuanlinggou Formation in North China", *Sci. Adv.* **10** (2024).

adhesion. Animals can be distinguished from the other multicellular lineages by having an epithelium, a single layer of tightly packed cells held together by cell junctions that have different functions within the epithelium. In particular, the "adherens junctions" tether adjacent cells to one another.

Animal cells do not have a cell wall, which allows them to make dynamic changes during development in their adhesive properties by regulating the expression of different adhesion molecules. Those multicellular lineages that evolved from unicellular ancestors having rigid cell walls cannot change their linkages with other cells. Adhesion is established as new cells form and remains the same throughout life. In this type of multicellular development cells divide and remain linked by their shared cell wall. Something similar to this would have to evolve on Earth-like planets.

Turning to symmetry, in biology it is always approximate despite the impression one may get from observation. Animals having bilateral symmetry are known as Bilaterians. In particular, their embryos have bilateral symmetry. The term Radiata applies to animals with radial symmetry.

Exact symmetry doesn't exist in animal body shapes since morphogenesis is regulated by mosaically acting gene regulatory networks. The difference between whole body symmetry and smaller anatomical structures depends on the timing during development. Holló[2] concludes that

> "The genome of most animals can express both radial and bilateral symmetries, which are the two possible local optima in the morphospace of theoretical body geometries, and are tightly linked to the function of the given structure. These notions together support the idea that radial and bilateral symmetries seem to be obligatory patterns in the evolution of animal body plans, and that the most important limits to the evolution of symmetry were — and are — ultimately settled by the physical environment of planet Earth."

This would presumably also hold for other Earth-like planets.

---

[2]Holló, G., "A new paradigm for animal symmetry", *Interface Focus* **5**(6) (2015), 20150032. doi: 10.1098/rsfs.2015.0032. PMID: 26640644; PMCID: PMC4633854.

Bilaterally symmetric animals are able to develop streamlined bodies that allow movement in a particular direction, which allows the possibility of having a head and a tail end. During motion, the head encounters the environment first so that sense organs, such as eyes and mouth, would generally be located there. Radially symmetric animals, such as sea anemones, jellyfish, and starfish, are usually sessile. This category includes animals that float and those that are slow-moving like starfish. The Bilateria represent some 90% of all animals.

How and when did these symmetries evolve? On Earth, the first known multicellular eukaryotes date back some 1.2 billion years, but more complex life did not begin until the Cambrian explosion which began about 542 million years ago. The "Cambrian explosion" refers to the time when bilaterally symmetrical animal groups diverged from a common ancestor during the early part of the Cambrian period. This is thought to have set the stage for the evolution of most of the diversity of animal life and led to the extant phyla. The major groups of bilateral animals have distinct body plans, each belonging to its own phylum. All multicellular animals except sponges and coelenterates have bodies that derive from three embryonic cell layers, the ectoderm, mesoderm, and endoderm. Extant echinoderms, such as sea stars, sea urchins, and sea cucumbers, appear to be radially symmetric, but their larvae are bilaterally symmetric. The Cambrian explosion corresponds to the transition between the Proterozoic and Phanerozoic eons.

The Cambrian explosion had two phases separated by the Sinsk Event, which occurred about 513 million years ago. The first phase was dominated by non-bilaterians such as Porifera, Cnidaria, and Ctenophora; the second by radiating non-bilaterian and bilaterian species.[3] The second phase may have been terminated by the late Cambrian SPICE event[4] (SPICE means

---

[3]Zhuravlev, A.Y. and Wood, R.A., "The two phases of the Cambrian explosion", *Nat. Sci. Rep.* **8** (2018), 16656. doi:10.1038/s41598-018-34962-y.

[4]https://en.wikipedia.org/wiki/Steptoean_positive_carbon_isotope_excursion.

Steptoean (late Cambrian) Positive Carbon Isotope Excursion),[5] which occurred some 495 million years ago and lasted for around 2–4 million years.

The rise of atmospheric oxygen and an increase in calcium concentration in the ocean are thought to be the main causes of the Cambrian explosion.

The oxygen concentration, given in percent, over the last billion years is shown in Figure 1.

Stratigraphic and geochemical data show that marine sediments from about 540 to 480 million years ago show that there was an expansion in the area of shallow epicontinental seas and an anomalous pattern of chemical sedimentation implying an increase in ocean alkalinity and enhanced chemical weathering of continental crust. The resulting globally occurring stratigraphic surface which in

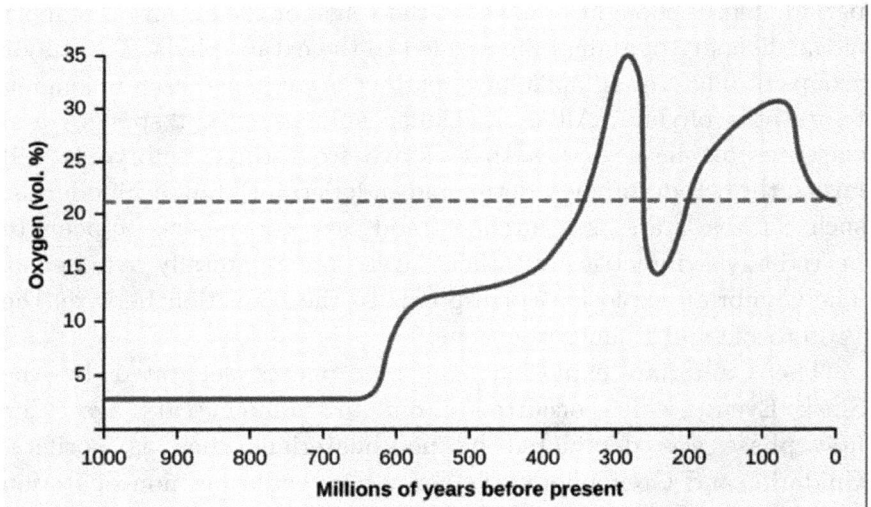

Figure 1. Oxygen content of the Earth's atmosphere over the last billion years [LibreTexts: The Cambrian Explosion of Animal Life: 27.4B].

---

[5]Kozik, N.P. *et al.*, "Protracted oxygenation across the Cambrian-Ordovician transition: A key initiator of the great ordovician biodiversification event?", *Geobiology* **00** (2023), 1–18.

most regions separates continental crystalline basement rock from the younger Cambrian shallow marine sedimentary deposits is known as the Great Unconformity. The formation of the Great Unconformity is thought to have been a cause for the evolution of biomineralization and the Cambrian Explosion.[6]

The recovery from "Snowball Earth", a period of global glaciation thought to have occurred before 650 million years ago during the Cryogenian period, may also have played a role in the Cambrian Explosion.

The current understanding of how bilateral symmetry evolved during the Cambrian Explosion is that it came from a bacterium whose genes had a "polarity", or a tendency to develop a distinct anterior and posterior, so as to grow or orientate in a particular direction. It has been hypothesized[7] that the evolution of polarities and the emergence of Bilaterians was induced by the oscillations in glycolysis, which itself was regulated by asymmetrical enzymes.

Glycolysis is the only pathway that can generate ATP in the absence of oxygen. Glycolysis is an ancient pathway that evolved before oxygen was present in the Earth's atmosphere and is conserved in living organisms (see Figure 1). In glycolysis one molecule of glucose is split into two molecules of pyruvate.[8]

Bilateral symmetry then appears to be a requirement for the evolution of higher-level life forms. The implication is that it would also be a requirement for higher-level life on all Earth-like planets as well.

---

[6]Peters, S.E. and Gaines, R.R., "Formation of the 'great unconformity' as a trigger for the Cambrian explosion", *Nature* **484** (2012), 363–366.

[7]Toxvaerd, S., "The emergence of the bilateral symmetry in animals: A review and a new hypothesis", *Symmetry* **13** (2021), 261.

[8]Chandel, N.S., "Glycolysis", *Cold Spring Harb Perspect Biol.* **13** (2021).

Part II

Intelligent Life on Earth-Like Planets

Chapter 7

Some Necessary Background: Origin of Intelligent Life on Earth

What is intelligence? How can it be measured? These are very difficult questions. Intelligence clearly has to do with the ability to think in terms of language and vision, to name a few modes of thought. One thing is clear, the basis for all modes of thought used by all animals including human beings can be traced back to their sensory perceptions.

Animal interpretation of the world around them can be very different from each other and from human perception. Local sensory neuronal networks associated with sensory neurons that detect signals from the surrounding environment often perform some processing of the received sensory data before the resulting coded information is sent on to the brain. The brain itself is organized around receiving, processing, and storing this information upon which responses are based.

The neuronal networks of the lower animals, such as jellyfish and some insects, are almost hard wired in the sense that little if any learning is required for their full behavioral repertoire to become available after they are fully formed. In the higher animals, much of this ability to process sensory information is developed during early life and is a learned process in that, while the basic architecture of the nervous system is genetically determined, its development depends on early sensory stimulation and is therefore a learned process in the broader meaning of the term.

In the case of human beings, intelligence cannot properly be encapsulated in a single factor like I.Q. Intelligence has a multi-dimensional nature, the components of which vary across populations. Whatever capability one chooses to measure (the most often mentioned being verbal and spatial abilities), the results of the measurements taken across populations will fall into roughly a bell-shaped curve known as a normal distribution, with the peak at the mean value. In a normal distribution, half the population will have a given capability above the mean and half will be below. People will have a range of capabilities, some above the mean and some below. Each capability will be due to a mixture of factors including genetic, developmental and cultural. Upbringing, nutrition (whether *in utero* or later), interaction — particularly at an early age — with parents and other adults, general and formal education, and the general physical and social environment are particularly important. In other words, the result of both nature and nurture.

To understand the difficulty in determining intelligence, consider the case of idiot savants, a term used to describe a person having exceptional capability in one mode of thought such as for example mathematics or music, but have significant impairment in other areas of intellectual or social functioning. When the term was first introduced by John Langdon Down in 1887, the term idiot meant having an I.Q. in the range of less than 20–25! Today, the less pejorative terms "savant syndrome" or "savant skills" are used. What one means by savant here is a skill so exceptionable that almost no non-disabled people can match it. How does one judge the intelligence of such people? Should the abilities and disabilities be averaged? None of the usual measures really make sense, so one usually sidesteps the issue and describes the syndrome. Another solution was found by Nettelbeck and Young[1]:

> "On the basis of recent research, two characteristics of savant performance are identified; the first is soundly functioning long-term memory that is narrowly focused and the second is a specific aptitude; that is, memory and cognitive processes dedicated to a specific ability. It is concluded that savant skills are not intelligent ...".

---

[1]Nettelbeck, T. and Young, R., "Intelligence and savant syndrome: Is the whole greater than the sum of the fragments?', *Intelligence* **22** (1996), 49–68.

It is doubtful that this conclusion will satisfy very many people.

Does the savant syndrome appear in other animals? In the end it is hard to know what intelligence really is since it depends on the behavioral context used to judge it.

Since behavior and intelligence are phenomena associated with the brain, it makes sense to try to compare the brains and intelligence of different animals. During the 20th century, and even earlier, many quantitative comparisons between animals were made of brain size and structure, and in particular of the cerebral cortex. As mentioned in the Preface, the human cerebral cortex is unique because it tripled in size over the last 1.5 million years. This is the source of the higher-level intelligence found in modern humans. Understanding how this happened is crucial for trying to ascertain whether such intelligence could arise on other Earth-like planets.

Truly intelligent life appears on Earth only among anthropoid primates and, as will be discussed later, the exceptional case of the Cetaceans. The anthropoid primates, like all primates, descended from tree-dwellers and consequently have hands, which, as will be discussed below, is one of the principal requirements for the evolution of higher-level intelligent life. The most important requirement is for a large number of neurons in the cerebral cortex, which would not occur without being able to meet the increasing energy requirements needed to support the number of neurons in the evolving cerebral cortex.

How many neurons are in the human cerebral cortex and how does this compare to the brains of those that are the ancestors of modern humans? This is not a simple question.

Until recently, comparing the brains of different species was limited to obtaining estimates of the density of neurons in sections of the tissue from the brains of those species. Using a different and innovative technique led Suzana Herculano-Houzel[2] to conclude that

---

[2]Herculano-Houzel, S., *The Human Advantage: A New Understanding of How Our Brain Became Remarkable* (The MIT Press, Cambridge, MA 2016), as well as some seventeen references to Herculano-Houzel's scientific papers given in the book.

there was not a universal relationship between the mass of brain structures and their number of neurons: because humans are primates their brains must only be compared to other primates. The result of doing this is shown in Figure 1. As can be seen, primate and non-primate brains scale differently.

In general, how does the human brain differ from that of other primates? In the past, the ratio of brain mass to body mass was the single most important factor to consider in comparing primates. This has always been somewhat suspect since there is no differentiation between neural density and neuron size across species, nor the number of neurons in each of the principal parts of the brain, essentially cerebral cortex and the cerebellum. This has now dramatically changed with the research done by Herculano-Houzel who argues that,

> "the human advantage lies, first, in the fact that we are primates, and, as such, owners of a brain that is built according to very economical scaling rules that make a large number of neurons fit into a relatively small volume, compared to other mammals."

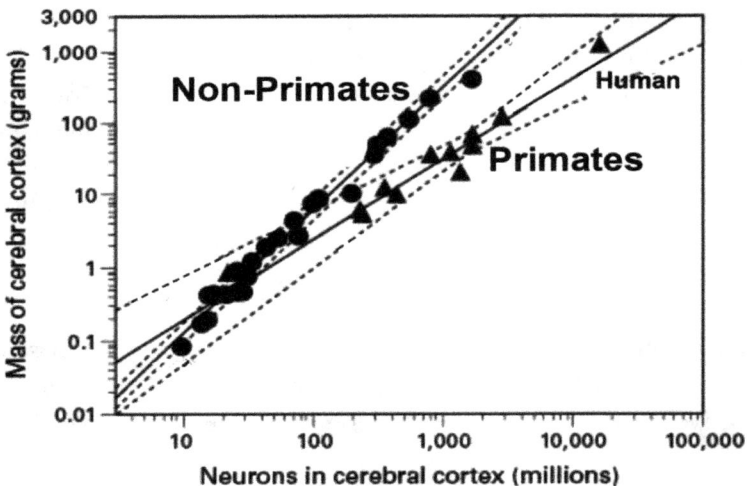

Figure 1. Mass in grams in the cerebral cortex of different generic non-primates (filled circles) and primates (filled triangles). The dashed lines indicate the 95% confidence intervals for each plot. Note that humans are contained within the primate confidence interval. [From Suzana Herculano-Houzel, *The Human Advantage*, Fig. 5.2].

The brain of modern humans uses about 500 Calories[3] a day, a quarter of the 2000 calories used by the entire body. The relatively low amount of caloric intake available from foraging limited the brain size of earlier forms of humans to perhaps 50 billion neurons. Herculano-Houzel found that despite higher estimates in the literature, modern humans have about 86 billion neurons, and that,

> "the human brain was simply a scaled-up primate brain, the mass of the human cerebral cortex was just what would be predicted for a generic primate brain having the same mass in its rest of brain."

The last part of this quote was presumably intended to mean that the mass of the human cerebral cortex is what would be expected for any primate brain having the same total mass. The relative brain mass of the cerebral cortex for various primates is shown in Figure 2(a).

Herculano-Houzel found that as the number of neurons in the brain increases, different scaling rules apply to the cerebral cortex and the cerebellum; the mass of the cerebral cortex increases faster than the cerebellum. This is because cerebral cortical neurons must connect over greater distances than those in the cerebellum — this adds additional mass to each neuron. Humans have the largest brain of all primates and consequently the largest cerebral cortex. The percentage of all neurons in the brains of mammals found in the cerebral cortex is about the same.

In Figure 2(b), the elephant corresponds to the filled square furthest to the right and is clearly an outlier. The elephant brain has around three times as many neurons as the human brain, but only about 2.2%, or 5.6 billion, of these are found in the elephant's 2.8-kg cerebral cortex and 97.5% are in the elephant's cerebellum; this is some ten times what would be expected. Humans have about 16 billion neurons in their 1.2-kg cerebral cortex, almost three times as many as in the elephant's cerebral cortex. The reasons for this anomaly are not yet understood, although there has been some

---

[3]This common usage of the term "Calorie" stands for kilocalorie or 1000 true calories of energy (in cgs units). One kilocalorie is the amount of energy needed to raise the temperature of a liter of water 1°C at sea level.

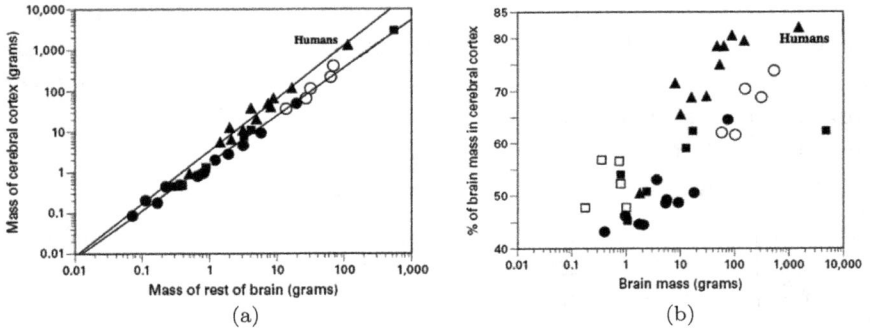

(a)   (b)

Figure 2. (a) As the mass of the brain increases, the mass of the cerebral cortex increases faster in mammals than other animals. Primates have the greatest rate of rise. For a given total brain mass, primates have a larger cerebral cortex than other animals; (b) for a given brain mass, the percentage of brain mass in the cerebral cortex differs among mammalian groups, with humans having the largest fraction. The open squares are for animals like hedgehogs, shrews, and moles — often referred to as insectivores; the filled squares are for mammals of African origin like elephants, sea cows, and several others; the fill circles represent rodents of similar brain mass as the latter two categories; open circles correspond to animals like cows, sheep, pigs or camels; and triangles represent primates. [Abstracted from Figs. 7.1 & 7.2 of *The Human Advantage: A new Understanding of How Our Brain Became Remarkable*].

speculation that it has to do with sensory processing of information from the elephant's trunk.

Here is a comparison of the cerebral cortex of humans and elephants: Primate neural density in the cerebral cortex is $\sim$40,000 neurons per milligram (n/mg) independent of the primate being considered, whereas neural density in non-primates decreases with an increase in the number of neurons in their cerebral cortex; e.g., an elephant has $\sim$5,000 n/mg with the number of neurons in the cerebral cortex being $\sim$5 $\times$ 10^9. Humans have $\sim$1.5 $\times$ 10^{10} neurons in a cerebral cortex weighing $\sim$1 kg. With the primate neural density given above, the number of neurons in the human cerebral cortex is $\sim$4 $\times$ 10^{10}, some ten times that of the elephant. Primates branched from the non-primate common ancestor so as to uncouple the addition of neurons to increases in the size of neurons in the cerebral cortex, which leads to lower neural densities.

With the evolutionary growth of the cerebral cortex in early humans came the ability to imagine, to have a sense of the past and future, and ultimately led to the ability to communicate and detect the mental states of others.

The neocortex of mammals is subdivided into functionally specialized zones. Mammals with small brains with a correspondingly small neocortex may have 20 or fewer zones, while primates range from 50 or less areas to 200 or more, per cerebral hemisphere, in humans.[4,5]

But it was ultimately the ability to remember incidents of damage and danger from fire, and nonetheless learn to control and use it, that led to the development of the modern human brain.

---

[4] J. H. Kaas, "The organization of neocortex in early mammals", In: S. Herculano-Houzel (ed.), *Evolution of Nervous Systems, Vol 2. Mammals*, 2nd edition (Elsevier, London 2017), pp. 87–101.

[5] Van Essen, D. C. and Glasser, M. F., "The human connectome project: Progress and prospects", *Cerebrum* (2016), 1–16.

Chapter 8

The Cultural Aspect of Human Physical Evolution

In 2009 Richard Wrangham[1] in his book *Catching Fire* introduced what became "the cooking hypothesis". In his words, what

> "gave rise to the genus Homo, one of the great transitions in the history of life, stemmed from the control of fire and the advent of cooked meals. Cooking increased the value of our food. It changed our bodies, our brains, our use of time, and our social lives. It made us into consumers of external energy and thereby created an organism with a new relationship to nature, dependent on fuel."

Few people have emphasized the importance of the invention of cooking, probably by *Homo erectus*, to the evolution of the human brain. The brain mass of the Homo lineage increased from less than 0.5 kg some 7 million years ago to about 0.75 kg about 1.5 million years ago. At that point it was probably *H. habilis* who introduced the cooking of food that led to a brain mass of about 1.5 kg in only 1.5 million years, a phenomenal growth in such a relatively short period of time. This is shown in Figure 1.

As can be seen from Figure 1, the brain mass of the pre-homo lineage had risen to about 0.75 kg, which is probably the most that could be reached by foraging for food. The invention of cooking dramatically lowered the foraging time needed to obtain the great increase in calories needed to allow the evolution of a brain mass of some 1.5 kg.

---

[1]Wrangham, R., *Catching Fire: How Cooking Made Us Human* (Basic Books. A Member of the Perseus Book Group, New York, 2009).

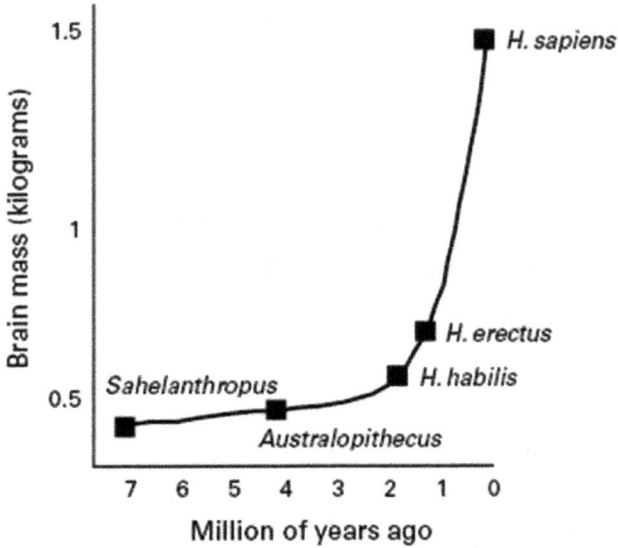

Figure 1. The dramatic rise in brain mass after the invention of cooking, probably by *H. Habilis*. [From Fig. 11.1 of *The Human Advantage: A new Understanding of How Our Brain Became Remarkable.*]

The use of the heat from fire to cook meat breaks down its collagen fibers making it softer and more palatable and in the case of plants makes the starch and fat available by softening the cell walls.

In summary: Because eating raw foods can yield as little as a third of their caloric content compared to close to a 100% for cooked foods, the ability to use fire to cook food post *Homo erectus*, was a requirement for the evolution of the modern human brain with its large number of neurons in the cerebral cortex; this in turn required that our ancestors be primates so that they could have hands to allow them to use fire.

It is interesting to note that the invention of cooking had not come about directly through Darwinian evolution but is rather the first clear evidence of what could be called cultural evolution.[2] Of course,

---

[2] "Darwinian evolution" is used here to mean the *Modern Synthesis*, which incorporated Mendel's discoveries. The fact that genes or their epigenetic

once this innovation occurred the following evolutionary changes were due to Darwinian evolution and the selective advantage of an increasingly larger brain mass and intelligence. The evolution of culture *per se* is discussed in the following chapter. An interesting discussion of the relationship between cultural and genetic evolution has been given by Creanza *et al.*[3] For a discussion of how culture can modify genetic selection see Whitehead *et al.*[4]

The classic example of the evolution of culture affecting genetic evolution is the continued production of lactase into adulthood allowing adults to digest lactose. This came from the development of cattle domestication and dairying. The role of epigenetic control of gene expression in such phenotypic modifications, can greatly accelerate adaptation to changing environmental conditions. In essence, epigenetic forms of inheritance can influence evolutionary trajectories. The biochemical methods involved have been discussed by Stajic and Jansen.[5] The molecular mechanisms of transgenerational epigenetic inheritance has been discussed by Fritz-James and Cavalli[6] and is available in the HAL open-access archive. What follows is a general an introduction to epigenetics.

Introduction to Epigenetics

Epigenetics should not be confused with phenotypic plasticity. Phenotypic plasticity is the ability of an individual genome to

---

regulation can evolve in a culturally determined environment should not be misinterpreted as justifying some form of the long-discredited idea of Social Darwinism. For an extensive discussion of these ideas see the section *Cultural Evolution and Social Darwinism* in my 2018 book *The Immense Journey.*

[3]Creanza, N., Kolodny, O. and Feldman, M.W., "Cultural evolutionary theory: How culture evolves and why it matters", *PNAS* **114** (2017), 7782–7789.

[4]Whitehead, H. *et al.*, "The reach of gene-culture coevolution in animals", *Nat. Commun.* **10** (2019), 2405.

[5]Stajic, D. and Jansen, L.E.T., "Empirical evidence for epigenetic inheritance driving evolutionary adaptation", *Philos. Trans. R. Soc.. Lond. B Biol. Sci.* **376**(1826) (2021), 20200121.

[6]Fritz-James, M.H. and Cavalli, G., "Molecular mechanisms of transgenerational epigenetic inheritance", *Nature Reviews Genetics* **23** (2022), 325–341. (https://hal.science/hal-03805066)

produce different phenotypes when exposed to a changing environment. It can affect physiology, morphology and behavior and is widespread in both plants and animals. An extreme form of phenotypic plasticity is the phase polymorphism exhibited by desert locusts when population density increases beyond a critical point. These insects exhibit reversible changes known as "phase changes". In the migratory locust there has been an analysis of gene expression associated with their phase change showing that rapid expressional changes occurred in 1,444 of the 9,154 genes analyzed. It has also been shown that in the desert locust serotonin levels play a major role. The change in behavior induced by crossing the critical population density is strongly correlated with an increase in serotonin.

Darwinian evolution maintains that inherited variations depend on genetic changes in the DNA base sequence. Epigenetics adds to this by recognizing that variations in the expression of a small number of genes controlled during development (ontogeny) by a regulatory network can have a dramatic impact on the phenotype. It should be remembered that DNA not only has genes that code for proteins, but also non-protein-coding genes and regulatory elements that control gene expression. How genes are regulated during ontogeny is what makes the greatest difference among animals. We share many genes in common with flies and worms. And with mice and chimps the percentage shared is over 95%!

Epigenetic changes are those that modify DNA and chromatin — when it is present — so as to change gene expression without changing the underlying DNA base sequence. The exact mechanism for transforming epigenetic changes into genetic changes remains uncertain. In bacterial experiments the most beneficial mutations, which contributed most to the early improvement in fitness, involved global regulatory functions. These kinds of mutations have a pleiotropic effect (one that affects several traits simultaneously) on the expression of many genes. Some of these changes will be beneficial and some will not. To contribute to the rapid early rise in fitness the changes must on the whole be beneficial. Natural selection would rapidly weed out the maladapted bacteria leaving only those mutations that increased fitness. Because many genes are affected in pleiotropic mutations, the rise in fitness could be significantly greater than the linear increase in fitness that would be expected from genomic evolution.

Epigenetic changes occur under strict control during the development and differentiation of the various cells in an organism. It is the changes in the control of gene expression that is responsible for the growth of different organs and tissues without changing the sequence of nucleotides in the DNA that is common to almost all cells. Subsequently, each type of cell expresses only a subset of the genes contained in the genome. In ontogeny the epigenetic program used is genetically controlled and as a result some have argued that epigenetic states are themselves genetically encoded.

Despite this strict control during ontogeny, environmental changes can not only induce changes in the regulation of genes, but such changes may be heritable. This is the way used by animals to quickly adapt to rapidly changing environmental conditions. The mechanisms behind the change in the regulation of genes involve the non-coding genes that produce RNA used in the regulatory process, DNA methylation — which adds a methyl group (a molecule having one carbon atom bonded to three hydrogen atoms) to cytosine, one of the four chemical bases that make up the genetic code — and chromatin remodeling.

DNA in the nucleus of a cell wraps around proteins call histones and the complex then coils up into a dense structure called chromatin. Methylation usually means the gene is inactivated, while chromatin remodeling involves modification of the histone proteins by adding methyl, phosphate, or acetyl groups. Genes that have acetylated histones are generally active. There is evidence that DNA methylation and histone modification act together to regulate gene expression although the mode of communication between the two is not yet fully understood.

This discussion has not yet addressed exactly how epigenetic changes are transformed into genetic changes. Although there are still many questions, there is some information available. The genetic code is written in terms of the sequence of the bases adenine (A), guanine (G), cytosine (C) and thymine (T) that appear in the DNA molecule. These bases generally appear as pairs of A-T and G-C along the length of the molecule. These bases generally appear as pairs of A-T and G-C along the length of the molecule. The ends of a DNA molecule are designated as $3'$ and $5'$; the molecule has two spiral strands where the $3'$ end of one is paired with the $5'$ end of the other. The acronym CpG stands for $5'$—C—*phosphate*—G—$3'$, where the phosphate links any two nucleosides together in the DNA molecule, here the nucleosides C and G. Now if a methyl group is attached to a cytosine appearing in a gene (the process is called methylation), it can affect the expression of that gene thereby becoming a component of the epigenetic regulation of the gene. The molecule that results from the methylation of cytosine is called 5-methyl cytosine.

5-Methyl cytosine has the property that it is susceptible to spontaneously deamination (the removal of an amine group — a negatively charged ion having one nitrogen and two hydrogen atoms) resulting in the base thymine giving the DNA molecule a G/T mismatch since G normally pairs with C not T. Should the molecule now replicate, the result would be a post-replicative mutation when the T pairs with an A at that site of the DNA molecule. The original C-G pairing has given rise to an A-T pairing in one of the molecules after replication. Single nucleotide polymorphisms (SNPs) often occur in animal genomes, the most common such polymorphism being the transition from a C to a T nucleotide. Methylated cytosine has a higher mutation rate, transforming it into thymine, than non-methylated bases. This hyper mutability provides a possible mechanism for transforming epigenetic changes to genetic ones.

The change in the genome by SNPs at CpG sites is due to G/T mismatches, which if detected before replication, are repaired by the enzyme thymine-DNA glycosylase (TDG). How successful this is will be dependent on the transcription of the gene for TDG. It has been shown that wild-type p53 (a protein whose structure has not been changed by a mutation of its gene) regulates TDG expression and a loss of p53 function will decrease genetic and epigenetic stability. (Unmutated p53 is known as the "Guardian of the Genome" and protects the genome under conditions of stress. As well as its role in the prevention of cancer, it plays a key role in the regulation of multiple cellular stress responses.)

• • •

Intelligent life on Earth-like planets would also be expected to have a relationship between cultural and genetic evolution. This almost goes without saying since the ability to use fire to cook food is a requirement for the evolution of a brain with a large enough number of neurons to form the equivalent of our cerebral cortex.

Chapter 9

The Late Human Advantage

What then gave modern humans the enormous advantage they have over other animals? It can't be simply from the large number of cerebral cortex neurons. That is no doubt necessary, but not sufficient. After all, Neanderthals' brain size was if anything slightly larger than ours and there is no reason to think that their cerebral cortex held significantly fewer neurons. The answer, as so strongly emphasized by Edelman,[1] is true language, but culture also plays an important role. Human culture evolved slowly in several stages. The use and making of tools are perhaps the oldest achievements. Various animals from birds to simians use tools of the simplest sort, such as a stick or rocks, and even pass specific skills on from generation to generation, but not at the level of even the earliest humans. The use of fire also appeared among early humans as well as the domestication of wolves presumably for hunting. But most important was the use of language. Some 50,000 to 80,000 years ago there was a flourishing of symbolic use that is believed to be a result of neural changes and reorganization in the brain that resulted in an enhanced ability to use language in a way that greatly improved communication. This created an enormous selective advantage that rapidly spread through the population. The result was the development of sophisticated hunting and gathering cultures that spread throughout the globe. Each showed variations in terms of tools, shelter, and other cultural adaptations to local conditions.

---

[1]Edelman, G.M., *Bright Air, Brilliant Fire: On the Matter of the Mind* (Basic Books. A Division of Harper Collins Publishers, Inc., New York, 1992).

Finding food and the activities associated with it left little time for any other activities, but nonetheless these societies developed extensive religious and oral traditions.

Language should not be thought of simply as communication by spoken words. All languages including tonal languages have emotional intonation. Lexical tones and intonation interact so as to express different emotions as well as the information contained in the speech. Different speakers show consistent sound production patterns.[2]

Unfortunately, language can be used not only as a mode of thinking and for enhanced communication, but also for darker demagogic purposes; emotions can be induced for the communication of powerful dreams that form the basis for real political actions.

There are also other modes of communication, such as mathematics, which is common in one form or another to almost all societies. While mathematical symbols are often given names, these names are a convenience and do not affect the actual nature of mathematical thinking; almost all of which is done visually and in terms of non-verbal concepts. Poetry is obviously verbal, and it is often used to induce visual and emotional states, sometimes having great insight and beauty, and sometimes to induce very dark emotions. And, of course, there is music, which has a similar effect and has its own written "language" that can convey pitch, rhythm, and has different forms to accommodate different instruments and modalities.

Human beings, along with many other animals, have a variety of senses including the obvious ones of sight, hearing, touch, smell, and taste. Some organisms have different suites of senses like the ability to sense and use electric or magnetic fields for practical purposes such as finding prey and navigation, or having extended, or at least different, visual or hearing ranges than human beings. This variation can be expected to be at least as great, and possibly greater, in intelligent creatures on Earth-like planets.

---

[2]Li, A., Fang, Q. and Dang, J., "Emotional intonation in a tone language: Experimental evidence from Chinese", *ICPhS XVII, Regular Session*, Hong Kong, August 2011, pp. 17–21.

We learn to interpret visual information as shapes, and we similarly learn to interpret a sequence of sounds as music or language. This interpretational process involves memory of previous exposure to similar sensory input. A given neural network can have many different patterns of excitation since individual neurons making up the network may or may not be an active part of any particular pattern of excitation; some are excited and some are inhibited. The number of possible combinations of even a few thousand neurons is enormous. This allows lower animals to display rather complex behavior with only a limited number of neurons.

Recent studies in humans show that the same neural representations are activated for perception and when viewing remembered visual images, and repetition improves the correlation. If we listen to a piece of music that has been previously heard, a memory of the piece is retrieved after hearing only the first few notes, and if an incorrect note is played, we notice it immediately. Similar behavior is displayed by other animals: a bird that has learned about a predator's shadow at an early age does not need to carefully categorize its image at a later appearance, it reacts instantly once enough information is available to call up the earlier memory; sometimes we see an animal on the lawn, only to have it almost instantly replaced by the crumpled paper bag it really is. Such misperceptions are common and clearly show that a camera is not a good model for vision; at best, the analogy serves to describe the projection of an image by the lens of the eye on the retina. The act of seeing itself involves visual processing by the retina and brain at the moment of seeing coupled with visual memories as imbedded in the neural networks of the visual cortex of the brain.

Among humans the ability to recall different types of sensory input is quite variable. Some people have superb visual memories both in space and time they are able to order them so as to dynamically visualize, for example, complex machinery and its operation. Some visualize in color, which itself can be variable since people's color vision differs, and others do not. Many cannot do either since they have very little spatial relations ability. Beethoven was surely able to recall music and "play it in his head"; after all, he was deaf when he wrote his late quartets, arguably his best works. While often being able to whistle a tune or recognize a piece of music, few people can actually "play" instruments in their head. But some can, and sometimes with great fidelity. Some can think using internal, silent words that they actually "hear" (known as sub-vocalization), others have only very limited ability to do so. Other variations may appear in the ability to recall tastes, smells, or other senses.

While the basic architecture of the nervous system is genetically determined, the information carrying capacity of DNA is far too small to specify the enormous number of interconnections of the brain. Instead, large quantities of excess neurons are produced during early life that forms the basis for later learning. Gerald Edelman[3] has developed the concept of neural Darwinism to help explain the selection of these neurons to form functional neural networks, and thus the micro-architecture of the brain. Learning involves the formation of memories; how these are formed and stored, a subject having a long history and literature, has now been understood at the neuronal and molecular level. It is well described in Eric Kandel's memoir and scientific exposition, *In Search of Memory.*

What appears to be the case then is this: Human beings think using a variety of basic elements related to our senses be they visual images, sub-vocalized words, or other recalled and composite constructs based on the modalities of our senses. Human beings use these elements of thought to enable what is called conceptual thought, that which involves abstraction and inference. It includes, for example, inductive and deductive logic, as well as mathematics and symbolic logic in its broadest sense. These elements can be used individually, and more often in combination. How people think of the same concepts, and which elements of thought are used and how they are combined, may vary from individual to individual.

What thinking may exist in non-human animals will reflect the elements of thought available to them from the neural cognates related to their often very different senses. This would be expected to be the same for intelligent life on Earth-like planets.

---

[3]Edelman, G.M., *Neural Darwinism: The Theory of Neuronal Group Selection* (Basic Books, Inc., New York 1987).

Do other animals on Earth communicate using language? Many communicate by sounds and some can recognize large numbers of words in human languages, but none appear to have true language.

Crows have been known to improvise tools, and myriads of animal owners have seen their pets respond to emergencies in ways that could neither be learned nor instinctive, but rather the result of mental problem solving. The lack of true language — with which higher-level consciousness in humans attains its maximal development — may not be necessary for the simpler forms of higher-level consciousness that appear in many animals. Even ducklings show some aspects of abstract thought in being able to understand the concepts of "same" and "different". Nonetheless, while the understanding and use of simple syntactical language comprised of up to three-word sentences is found in great apes, complex syntactical language is only found in humans. Dolphins are a special case.

Dolphins and the Possibility of High-level Intelligence on Earth-like Planets

Earth has two species with a high level of intelligence, humans and dolphins. As put by Marino,

> "The study of cetacean brains has revealed that the human brain is not the only brain that has undergone significant increases in size and complexity. Cetacean brains have, as well, but have done so by taking a very different neuroanatomical path to complexity, and although not the only one, they may be the most compelling example of an alternate route to intelligence on par with that of humans"[4].

As will be discussed below, the level of dolphin intelligence is very difficult to determine. Nonetheless, it is now well known that they are far more intelligent than thought of in the past.

Understanding the issue of language use and the nature of the high intelligence found in dolphins is a worthwhile exercise in the context of trying to explore the possibility of high-level intelligence

---

[4]Marino, L., "The brain: Evolution structure and function", in *Dolphin Communication and Cognition*, eds. Herzing, D.L. and Johnson, C.M. (MIT Press, Cambridge Massachusetts, 2015), Chapter 1.

on Earth-like planets. It also shows how the constraints of the environment may affect language structure and the multifaceted form that intelligence may take. For these reasons, what is known about dolphins will be gone into in some detail below. This covers their physiology, intelligence, language, and culture.

An analogy to the dolphin would not apply to Earth-like planets without continents since, as will be seen, high level intelligence would not be likely to evolve on such "water worlds".

$$\bullet\ \bullet\ \bullet$$

Dolphins use sound for echolocation and communication, but in the ocean sound propagation is not simple. Cetacean hearing and neural processing must deal with the curved ray paths of sound in the ocean, multipath, and the non-spectral information contained in echo signals. From such data dolphins may well be able to construct three-dimensional "images" of the objects they echolocate.

Dolphins are highly intelligent creatures perfectly adapted to their environment. Consequently, understanding the extent and nature of their intelligence is difficult. One big difference between dolphins and humans is that we have hands, allowing us to adapt to and change a variety of environments, and our cultural level can be measured by the use of tools and later, after the invention of writing, and by surviving texts in one form or another. Comparison of dolphin "culture" to that of humans is only reasonable if the comparison is made to the time when humans were still in the hunting and gathering phase, before the invention of agriculture allowed fixed settlements and the rise of civilization, and when they were living in the environment within which they evolved. Even if properly contrasted, given the very different environments it is unclear what measures could be used for comparison.

Like humans, the learning of each generation of dolphins would be lost to succeeding generations were it not for culture as a means of passing down that knowledge. Teaching by example is known to occur in dolphins and perhaps the best-known example is the use of sponges by a small fraction of the population in Shark Bay in western Australia.[5] Foraging and social behavior are taught to calves during their first few years by their mothers. Those mothers that use sponges teach their calves (mostly females) to use conical basket sponges to protect their rostrum when foraging on the bottom. Many other examples of teaching by example exist, but this is the only well-known one where tools are used by wild dolphins. The activity is advantageous to the dolphins because fish found in this way are highly nutritious. Another spontaneous example of tool use is discussed below. If dolphins have a true language there could also be an oral tradition, as was the case with humans up to the time writing was developed.

Is the use of the term "culture" with regard to dolphins justified? As put by Toshio Kasuya, upon receiving an award from the Society for Marine Mammalogy,

> "... culture exists if a cetacean species exhibits a behavioral trait that suggests culture, or if it has life history, or social structure that is suitable to maintain a culture, such as the short-finned pilot whales and some other toothed whales."

Much of the behavior of dolphins and whales is social and comes from being part of a community and therefore could be called cultural — it is very different from the behavior of a school of fish. They can imitate both human and dolphin behavior. An amazing example of this has been given in the book by Hal Whitehead and Luke Rendell titled *The Cultural Lives of Whales and Dolphins* published by the University of Chicago Press in 2015. The scientists involved observed the behavior of two female and one male bottlenose dolphin. During the period of observation, a human diver entered the tank where they were being kept in order to clean algae from the viewing ports. One of the dolphins

---

[5]M. Krützen *et al.*, "Cultural transmission of tool use in bottlenose dolphins", *PNAS* **102** (2005), 8939–8943.

"was then observed using a seagull feather he had recovered to stroke the viewing glass, apparently copying the cleaning (needless to say, this kind of housekeeping has never been observed in wild dolphins). He then apparently became quite taken with this. He went on to use a variety of objects to fulfill his role as everyone's favorite tank mate, including stones, paper, and even the fish he had been given to eat. He remained quite faithful to his human 'demonstrator' though, even mimicking the human divers' technique of holding onto steel bars beside the window to steady themselves, by placing a flipper in the same spot. He also became quite possessive of the viewing port and would aggressively prevent divers, and the other dolphins, from approaching it for a period of over fifty days. During this time though, he kept the window quite clean."

An even more astonishing imitation occurred when the calf of one of the females was observing the observer through the observation port. At the end of this mutual observation period, the human observer blew cigarette smoke at the port. The calf then "swam to her mother, from whom she was still nursing, and returned with a mouthful of milk and spat it back toward the window, producing an uncanny replication of a cloud of smoke. This apparently went on to become a regular trick." The young, nursing dolphin, who no doubt remembered that milk would make a cloud in the water when ceasing to nurse, had to conceive of the idea of using that observation to reproduce the behavior of the human and do it at the port rather than elsewhere in the tank. You blow smoke at me; I blow the equivalent of smoke at you! Dolphins also have a history of cooperating with humans and in particular with fishermen. A very interesting and popular article on interactions between cetaceans and humans was published by Charles Siebert in the 12 July 2009 issue of the *New York Times Magazine*.

Returning to Toshio Kasuya,

"... it is my view that cetaceans must depend upon knowledge accumulated through past experience. Such knowledge is likely to be transmitted, by learning, to other members of the group. This is the culture of the community."

The book by Whitehead and Rendell[6] is one of the few, if not only, generally readable scientific assessment of dolphin and whale social structure and culture.

Cultural learning occurs when some individuals learn a specific behavior from other members of their social group and then spread it to other members of the group. Above, the example of the use of sponges in foraging was discussed for dolphins in Shark Bay (Figure 1), but an amusing example is the fad of tail walking where a dolphin (Billie) learned tail walking on the surface of the water during a short period in captivity in a dolphinarium from performing

Figure 1. An adult female Indo-Pacific dolphin named Wave tail walking some seven years after Billie's 1988 release. [From Fig. 2 of Bossley *et al.*].

---

[6]Whitehead, H. and Rendell, L., *The Cultural Lives of Whales and Dolphins* (University of Chicago Press, Chicago 2015).

dolphins and was then released into the wild. Soon other dolphins in a local wild community began copying the behavior after observing Billie exhibit it. The fad soon disappeared from the community.[7]

What about language? Many animals communicate with each other, but the key elements of true language are semantics and syntax. Semantics roughly has to do with meaning and syntax with the rules for arranging words and phrases to produce well-formed sentences. Herman *et al.*,[8] used word order as a measure of syntax and found that dolphins are capable of understanding that changes in word order reflect changes in meaning. As put by him, one dolphin "was able to spontaneously understand logical extensions of a syntactic rule and was able to extract a semantically and syntactically correct sequence from a longer anomalous sequence of language gestures given by a human". Note the use of the word "gestures". Most dolphin communication with humans is based on the use of gestures and the dolphin's love of imitation. But this is very limited and it is suggested here that the use of a unique set of "whistles" having varying frequency and amplitude might be more appropriate. This to some extent done by Herman, Richards, and Wolz.[9] Their study was designed to determine if bottlenose dolphins could understand imperative sentences expressed in an artificial language. One of their dolphins was taught to recognize computer generated sounds. This part of their study addressed only the comprehension of the acoustic signals used and not the dolphin's ability to reproduce them. Their conclusion was that bottlenose dolphins can understand imperative sentences in terms of both syntax and the semantic components of the sentences using either the

---

[7]Bossley, M. *et al.*, "Tail walking in a bottlenose dolphin community: the rise and fall of an arbitrary cultur'l "a'' ", rsbl.royalsocietypublishing.org *Biology Letters* **14** (2018), 20180314.

[8]L.M. Herma", "Exploring the cognitive world of the bottlenose dolphin", in *The Cognitive Anima: Empirical and Theoretical Perspectives on Animal Cognition*, eds. Bekoff, M., Allen, C. and Burghardt, G. M. (The MIT Press, Massachusetts 2002), Chapter 34.

[9]L.M. Herman, D.G. Richards and J.P. Wolz, "Comprehension of sentences by bottlenose dolphins", *Cognition*, **16** (1984), 129–219.

artificial language based on acoustic sounds or one that was visually-based.

Ryabov[10] advanced the possibility that an interchange of acoustical signals recorded between two quasi-stationary Black Sea bottlenose dolphins constituted a verbal exchange. During the acoustic recording intervals, no special training or food reward was given to the dolphins. The recordings showed that the dolphins took turns in emitting their acoustic signals and did not interrupt each other, showing that each of the dolphins listened to the other's emissions before producing its own response. These responses were not repeats, and the alternate emissions from the dolphins differed in both their spectral components and length.

Ryabov observed that the recorded exchange

"resembles a conversation between two people. . . .Since the spoken language of the dolphin consists of spectral extrema that act as phonemes, we can hypothesize that it has both phonological and grammatical structures, so dolphins can create an infinite number of words from a finite number of spectral extrema, which can in turn create an infinite number of sentences. The analysis of the dolphin spoken language in this study has revealed that it either directly or indirectly possesses all the known design features of the human spoken language."

If this study is confirmed, it will place a special ethical and moral imperative on humanity to protect these creatures as intelligent beings.

Introduction to the Physiology of the Bottlenose Dolphin

A bottlenose dolphin can hear sound in the frequency range of ~75 Hz to ~150 kHz, compared to humans whose range is ~20 Hz to ~20 kHz. Dolphins are most sensitive to sounds between about 15 kHz and 110 kHz. As a result, the dolphin recordings in the

---

[10]Ryabov, V.A., "The study of acoustic signals and the supposed spoken language of the dolphins", *St. Petersburg Polytechnical University Journal: Physics and Mathematics* **2** (2016), 231–239.

general literature do not cover the frequency range of the majority of signals produced by dolphins. It should be noted that the higher frequencies have a much higher absorption coefficient in seawater ($\sim$18.3 dB/km at 100 kHz compared to 1.8 dB/km at 20 kHz), so that the higher frequencies are used for relatively local purposes such as echo location. It is now known that dolphins use echolocation to determine the shape of objects and that aural sensory information is integrated with visual information so that mental representations of shape are derived from both sources.[11]

Because dolphins are so superbly adapted to the environment within which they live, how to distinguish learned and instinctual behavior is not always apparent. It is also difficult to use anatomical information to determine the relative intelligence of dolphins. The neural connectivity patterns of their brains are very different from ours indicating that human and dolphin brains evolved along alternate paths to achieve the neurological and behavioral complexity exhibited by both. As put by Marc Lammers and Julie Oswald,

> "...less quantifiable, but perhaps most challenging, are the hurdles we must overcome related to the ecological, behavioral, and cognitive differences that exist between humans and dolphins. Although we may share certain similarities as a result of being social mammals, humans and dolphins occupy very different ecological niches and live in distinct sensory worlds. Therefore, we have to assume that substantial differences exist in the way humans and dolphins perceive and communicate about their world."[12]

In terms of the number of neurons in their brains, dolphins bracket the human brain; some having somewhat less than humans and

---

[11] While the concentration here is on the audition of dolphins, they also have other senses, among them an unusual form of vision where their eyes can be moved independently of each other and have good vision above and below the water surface. A review of dolphin sensory perception has been given by Kremers, D. *et al.*, *Frontiers in Ecology and Evolution* **4**, Article 49 (May 2016).

[12] Lammers, M.O. and Oswald, J.N., "Analyzing the acoustic communication of dolphins", in *Dolphin Communication and Cognition: Past, Present, and Future*, eds., D.L. Herzing and C. M. Johnson (MIT Press, Cambridge, Massachusetts, 2015), Chapter 5.

one species — the oceanic dolphin known as the long-finned pilot whale — has more than twice as many.

Both dolphins and whales form cultural communities, exhibit self-awareness, and are capable of planning ahead and understand linguistic syntax. All of which indicates that they are capable of abstract thought. One of the key questions is that while acoustic signals produced by dolphins represent a form of communication far more complex than the minimal form of language exhibited by many other animals, is it in some sense comparable to human speech?

Dolphins produce a variety of sounds and their "whistles" have been much studied. Each dolphin has what is known as a "signature whistle" that identifies it to other dolphins. Whistles are usually continuous and have a duration of tens of milliseconds to several seconds. These are narrow band signals whose frequency, usually between 2 kHz and 20 kHz, varies in time. They can be monotonic or have multiple inflections or steps in their contours. They may also have several harmonics that extend into frequencies greater than 20 kHz. Dolphins are also able to discriminate between whistles by their harmonic content.

There are two anatomical sites for sound production in the dolphin that can be controlled separately so that "whistles" and other sounds can be produced simultaneously. Some of these are high pitched clicks — whose frequency can exceed 100 kHz — that are generally used for echolocation, and various other types of pulsed sounds whose meanings are unclear. It should be emphasized that the signature whistles of bottlenose dolphins, with a frequency and harmonic contour that is unique to the individual, is crucial because it allows them to specifically address each other. Without this capability, their language would be limited to group communication signals such as "danger", "fish", "follow", "help", similar to what is exhibited by primates and birds.

Neurophysiology

Dolphins process auditory information in at least two areas of their brains — one adjacent to the primary visual cortex and one in the temporal lobe. The external part of the ear, the pinnae, and the external auditory canals were lost over evolutionary time. The lower jaw became the primary channel for sound. The tympanic membrane (eardrum) was replaced with a thin and large tympanic bone plate. The cochlea, while maintaining its basic structural form and function has also evolved to become better adapted to underwater existence. In essence, dolphins rely on sound conduction through a special "acoustic fat" channel in the lower jaw to carry sound directly to the bony case of the inner ear. Thus, the tympanic membrane and its path to the middle ear bones are largely bypassed.

In humans, the hair cells and auditory nerve fibers from the cochlea are limited in the frequency they can respectively generate and carry in response to an audio signal, a frequency ~ 3 kHz being the upper limit. For frequencies above ~ 3 kHz, the brain relies on the tonotopically organized basilar membrane of the cochlea, which provides a spatial separation of higher frequencies. Separate nerve fibers from the different tonotopic regions of the cochlea then carry information to the (secondary) auditory cortex of the brain to form a tonotopic map where different frequencies go to adjacent regions of the cortex. While the details will certainly differ, a similar solution to the frequency limits of nerve fibers (tonotopic brain maps) must exist in dolphins.

The common dolphin is smaller than the bottlenose dolphin and has a brain weighing about 800 g compared to 1500 g for the bottlenose dolphin. Humans have brains weighing ~ 1400 g.

Perhaps surprisingly, the very high frequencies that dolphins hear can also can be heard by humans. SCUBA divers can hear ultrasonic frequencies to greater than 100 kHz, but have no pitch discrimination above ~ 20 kHz. The evolution of the dolphin auditory system has resulted in greater high-frequency hearing sensitivity and very complex auditory processing.

Individual auditory nerve fibers of the human or dolphin transfer information from only a narrow part of the audible frequency spectrum. Electrophysiological recordings of the threshold response of the nerve fibers to sound are known as tuning curves (see Appendix C). They are plotted as the threshold intensity in dB required to achieve a response above the spontaneous firing rate of the associated neurons (in humans from essentially zero to 120 spikes/s) as a function of frequency. Tuning curves for the dolphin have been obtained by monitoring their auditory brain stem response, and are similar in shape to those seen in other mammalian species.

The vestibulocochlear nerve, known as the eighth cranial nerve, transmits sound and equilibrium information from the inner ear to the brain. The dolphin auditory nerve has several times as many fibers as the human eighth nerve, and the fiber diameters are also about twice as large as in humans, which about doubles their speed of signal propagation. The auditory tonotopic map in the dolphin brain has been displaced from the temporal to the parietal lobe (above the temporal lobe and behind the frontal lobe) and dorsal part of the hemisphere.[13]

In humans, as pointed out by Vanthornhout et al.[14] and R.V. Shannon et al.,[15] the primary factor in speech intelligibility is the temporal envelope having a modulation frequency below 20 Hz. The most important frequency range is 4–8 Hz corresponding to the average frequency of speech.

The auditory cortex of non-human primates is composed of areas known as the core, belt, and parabelt regions. In humans, the primary auditory cortex corresponds to the core regions. Kubanek et al.[16] found that the human non-primary auditory cortex faithfully tracks the speech envelope, a phenomenon often called cortical entrainment or phase-locking.

Experiments with bottlenose dolphins show that 14 ms tone bursts, amplitude modulated at 600 Hz, evoked a strong auditory

---

[13]S.H. Ridgway, "The auditory central nervous system of dolphins", in *Hearing by Whales and Dolphins*, eds. Au, W.W.L., Popper, A.N. and Fay, R.R. (Springer-Science+Business Media, LLC, 2000).

[14]J. Vanthornhout et al., "Speech Intelligibility predicted from entrainment of the speech envelope", *J. Assoc. Res. Otolaryngol*, **19**(2) (2018), 181–191. doi:10.1007/s10162-018-0654-z.

[15]Shannon, R.V. et al., "Speech recognition with primarily temporal cues", *Science* **270** (1995), 303–304.

[16]Kuganek, J. et al., "The tracking of speech envelope in the human cortex", *PLOS ONE* **8**, e53398.

neural response that is phase-locked with the envelope of the sound. The modulation index used in the experiment was unity (100% modulation), meaning that the amplitude of the carrier wave was the same as that of the modulating wave. The neural response was monitored as is done when recording an EEG: A recording suction cup electrode was placed on the dolphin's skin about 6 cm behind the blowhole with a ground electrode behind that followed by a reference electrode.[17] That is, the "recording electrode" and the "reference electrode" go to the inputs of a differential amplifier so that in-phase inputs cancel and out-of-phase inputs are amplified with respect to the grounded electrode. This allows signal amplitudes in the microvolt range to be recorded.

Dolphins can detect audio signals with a very fine time resolution, their integration times for a single short signal being on the order of 200–300 μs. Integration time is a measure of the ability to isolate individual acoustic events. A longer integration time corresponds to a reduction in temporal resolution. This fine time resolution allows dolphins to use high frequency broadband clicks of microsecond duration to visualize objects around them in terms of distance, shape, orientation, and composition. This is because the echo signal is amplitude modulated and several milliseconds in duration containing a great deal of non-spectral information.

The ocean is a complex medium where temperature, density, sound velocity, and salinity vary with depth. There are many factors that affect the return echo from a dolphin's clicks. Sound does not travel in a straight line in the ocean. For example, the mixed isothermal layer, which extends to about 60 m, forms a sound channel where sound emitted at a shallow angle returns to the surface where it is again reflected; the return signal from moving object would also be doppler shifted. Over short distances the curved ray paths would not matter, but for communication over longer distances it would. Multipath, where there is more than one propagation path

---

[17]Cook, M. L. H., "Behavioral and auditory evoked potential (AEP) hearing measurements in odontocete cetaceans",Graduate Theses and Dissertations (2006). https://scholarcommons.usf.edu/etd/2489.

for sound between the dolphin and the sound scattering object, is an important factor in spreading out the return signal. In shallow water, the dolphin must also take into account scattering from the seabed.

More discussion of human and dolphin hearing is given in Appendix C.

Chapter 10

Requirements for the Evolution of Intelligent Life on Earth-like Planets

Perhaps the best definition of Earth-like planets is that they have a surface temperature that allows liquid water to exist. The crust of the planet, if it is similar to the Earth would be mostly solid rock composed of mostly basalt and granite. If the planet has continents and is similar to the Earth, the continental crust would be mainly composed of granite whereas the oceanic crust would be denser and thinner than the continental crust and primarily composed of basalt.

In addition, liquid water would have to continue to exist over billions of years. For example, it took about 500 million years for life to come into existence on the Earth. Mars had liquid water on its surface for a similar period of time starting about 4 billion years ago, but then gradually lost its surface water as its atmosphere became too cold and thin to maintain its surface water.[1] If life did come into existence during the period when Mars had liquid water on its surface it would almost certainly not have had enough time to evolve into complex life.

During the Hadean Eon, between 4.3 and 3.8 billion years ago, Earth's ocean carbonate geochemistry significantly changed. During the early Hadean seawater pH was probably similar to 5.8, and by the late Hadean seawater pH went up to almost neutral being probably ~6.8. This means that the carbonate chemistry of seawater in the

---

[1]Schmidt, F. *et al.*, "Circumpolar ocean stability on Mars 3 Gy ago", *PNAS* **119** (2022), e2112930118.

late Hadean-early Archean was similar to that of modern oceans.[2] Life then began in an ocean having carbonate chemistry similar to today's oceans.

Here is the critical assumption for what follows: The brain of modern humans uses some 500 kilocalories per day leading to a total energy cost for the body of ~2000 kilocalories per day. The estimate for the average energy cost for neural brain tissue is 6 kilocalories per billion neurons per day. This holds for all primate species. It will be assumed here that the analog for neurons of intelligent life on Earth-like planets will have energy costs similar to that of the modern human brain.

This is not an unreasonable assumption. On Earth, the first evidence that neuronal tissue had come into existence comes from fossils dated to some 525 million years ago.[3] Since then, there has been more than enough time for energy costs to be evolutionarily minimized so that it is doubtful that on Earth-like planets the analog of neurons would have energy costs less than that of the modern human brain. Similarly, as shown in Figure 1 of Chapter 8 for the homo lineage, it is unlikely that the brain mass of intelligent creatures on Earth-like planets could rise to more than ~0.75 kg if they relied on foraging for food without cooking.

Could the metabolism of creatures on Earth-like planets be more efficient than those on Earth so that cooking would not be required? This is very unlikely. The evolution of metabolism on Earth is known to have some dependence on self-organization into chemical cycles promoted by the flow of energy through the biosphere.[4] Conditions, and the role of self-organization, can be expected to be similar on Earth-like planets. Metabolism is an efficient, interconnected and

---

[2]Morse, J.W. and Mackenzie, F.T., "Hadean ocean carbonate chemistry", *Aquatic Geochem.* **4** (1998), 301–319.

[3]Kristan, Jr., W.B., "Early evolution of neurons", *Curr. Biol.* **26** (2016), R937–R980.

[4]Braakman, R. *et al.*, "Metabolic evolution and the self-organization of ecosystems", (2017), E3091–E3100.doi/10.1073/pnas.1619573114 (Published online).

ancient biochemical system, whose evolution has been extensively covered by Scossa[5] and Ralser.[6]

Metabolic cycles such as the Krebs citric acid cycle[7] have arisen over the course of evolution. A metabolic cycle is a process where an overall chemical change is brought about by a cyclic reaction sequence. The Krebs cycle has the advantage of being about twice as efficient as any alternative cycles, and is thought to have evolved from pathways for amino acid biosynthesis.[8] It is found in all living cells today. Compared to alternative cycles it has the least number of chemical steps and the greatest yield for the production of adenosine tri-phosphate (ATP). ATP is used to store energy at the cellular level by all extant life forms. The Krebs cycle has been shown to occur spontaneously (in reverse) under prebiotic conditions.[9,10]

Given the evolutionary optimization of the citric acid cycle it is very likely the same cycle would evolve on Earth-like planets.

The basics of metabolism are given in Appendix B.

Earth-like Planets Without Continents

At first one might think that not having continents would not affect the possibility of high intelligence on such worlds. This is because

---

[5]Scossa, F. and Fernie, A.R., "The evolution of metabolism: How to test evolutionary hypotheses at the genomic level", *Comput. Struct. Biotechnol. J.* **18** (2020), 482–500.

[6]Ralser, M. *et al.*, "The evolution of the metabolic network over long timelines", *Curr. Opin. Syst. Biol.* **28** (100402) (2021), 1–8.

[7]Baldwin, J.E. and Krebs, H., "The evolution of metabolic cycles", *Nature* **291** (1981), 381.

[8]Meléndez-Hevia, E., Waddell, T.G. and Cascante, M., "The puzzle of the krebs citric acid cycle: Assembling the pieces of chemically feasible reactions, and opportunism in the design of metabolic pathways during evolution", *J. Mol. Evol.* **43** (1996), pp. 293–303.

[9]Muchowska, K.B. *et al.*, "Metals promote sequences of the reverse Krebs cycle", *Nat Ecol Evol.* **1** (2017), 1716–1721.

[10]Zhang, X.V. and Martin, S.T., "Driving parts of krebs cycle in reverse through mineral photochemistry", *J. Am. Chem. Soc.* **128** (2006), 16032–16033.

the cetaceans on Earth have a relatively high intelligence far beyond other ocean creatures. But they did not originate in the ocean.[11]

They originated ~50 million years ago in the Eocene epoch. Early cetaceans were amphibious and were descendent from terrestrial *Pakicetus*. It is their ear bones that were used to link these creatures that lived some 50 million years ago to the evolutionary lineage of cetaceans. Thus, they originated from land mammals, and cannot be used as an example for the evolution of high-intelligence in the oceans of Earth-like planets without continents.

The cephalopods in Earth's oceans, and in particular the octopus, could well have analogs in the oceans of Earth-like planets without continents. All cephalopods have either limbs or tentacles, the difference being whether they have suction cups along the entire length of the limb or only near the end of the limb. The octopus's eight arms each have two rows of suction cups along their entire length. They also have three hearts and nine brains, a central brain and one in each arm. They have ~500 million neurons, two-thirds of which are distributed amongst their limbs. The limbs have a neural ring that allows information to pass between them bypassing the central brain. Also, they only live ~2–5 years. This is thought to be a consequence of a reproductive strategy where they breed only once in their lifetime. The actual mechanism has been explained by Wang *et al.*[12]

The giant Pacific octopus can solve puzzles, use tools and even mimic other species. Their intelligence is similar to crows and apes. They can also communicate visually, although not much is known about this ability. The larger Pacific Striped octopus also exhibits social behavior.

A full-grown giant Pacific octopus can weigh more than 50–100 pounds and can consume 2–4% of its body weight in a day. For a 100-pound body weight and a diet of raw sea food this translates

---

[11]Thewissen, J.G.M., Z.Y. *et al.*, "From land to water: The origin of whales, dolphins, and porpoises", *Evo Edu Outreeach*, **2** (2009), 272–288.

[12]Wang *et al.*, "Steroid hormones of the octopus self-destruct system," *Curr. Biol.* **32** (2022), 2572–2579.

to ∼650 calories a day. Far less than what is required for high intelligence comparable to that of a human.

One might ask: What about the dolphins? They forage for food and are yet able to support the energy demands of a large brain. Isn't this a counterexample? Again, the dolphins are a special case.

The resting metabolic rate of humans is about 3.5 ml of O_2 per kilogram of body mass per minute. The bottle nosed dolphin has a resting metabolic rate in the range of 0.76–9.45 ml of O_2 per minute per kg of body weight.[13] The resting metabolic rate of humans falls within this range. How does the dolphin support this by foraging? Wild dolphins with an average weight of 200 kg need some 16,500 to 33,000 calories per day, equivalent to some 10–25 kg of fish each day. To support these requirements, dolphin metabolism has evolved over time.

The normal dolphin diet is similar to the ketogenic diet in humans where a very low amount of carbohydrates is consumed being replaced by burning fat for energy. However, having a whole fish diet only yields amino acids so that the metabolic effect is primarily glucogenic where a pyruvate residue is produced in metabolism, which is then converted to a carbohydrate (as glucose) and finally stored as the complex carbohydrate glycogen.

Dolphin glucose tolerance curves are prolonged and their glucose blood values are very high and they show no signs of diabetes mellitus. As described by Ridgway,[14] the physiological picture of dolphins resembles

> "what had been described for hyperthyroid diabetics. Dolphins have elevated thyroid hormone turnover, and fasting dolphins maintain a relatively high level of plasma glucose. After dolphins ingest glucose, plasma levels remain high for many hours. Interestingly, plasma glucose must exceed 300 mg/dl (about twice as high as the human threshold) before glucose appears in urine."

---

[13]Fahlman, A. *et al.*, "Field energetics and lung function in wild bottlenose dolphins, Tursiops truncatus, in Sarasota Bay Florida", *Roy. Soc. Open Sci.* **5** (2018), 171280. www.rsos.royalsocietypublishing.org.

[14]Ridgway, S.H., *Front. Endocrinol.* **4** (2013), Article 152.

What this shows is an amazing adaptation over time by the warm-blooded land-based ancestor of the dolphins to life in the ocean. Such an adaptation to ocean life is only shown by the cetaceans. Life that evolved in the oceans would not have to go through the adaptation of starting with the metabolism of the warm-blooded land-based ancestor of the dolphins. They would evolve a metabolism suited to the ocean.

Most animal life is ectothermic with only birds and mammals being endothermic. The body temperature of ectotherms, the pace of their biochemical reactions, and the rate of their physiological processes are determined by the environment. The upper limit of metabolic performance of an aquatic ectotherm is about 35% of that of a mammal of the same size. For a full discussion of these issues see van de Pol *et al.*[15]

The clear implication of the above is that Earth-like planets without continents would not be able evolve high-level intelligent life.

Earth-like Planets with Continents

Some 600 million years ago the continents of Earth were inhospitable to life with only some photosynthetic bacteria, fungi, and extremophilic algae able to survive under the extant conditions. Only when fungi and algae began forming mycorrhizal relationships did life begin to flourish on the continents. In modern times 90% or so of all plant species depend on such relationships. One branch of the fungi contains the largest known living organism on Earth. It is a contiguous colony of the Armillaria genus in the Oregon Malheur National Forest, which covers over 900 hectares and has an estimated age of some 8600 years. The fungi are among the largest and most widely distributed group of living organisms. For more about fungi see the highly readable book by Sheldrake.[16]

---

[15] van de Pol, I., Flik, G. and Gorissen, M., "Comparative physiology of energy metabolism: Fishing for endocrine signals in the early vertebrate pool", *Front. Endocrinol.* **8** (2017).

[16] Sheldrake, M., *Entangled Life* (Random House, New York, 2020).

The most important event for life on Earth subsequent to its origin was the development of photosynthesis by cyanobacteria that appeared on Earth around 3 billion years ago; life could not flourish before significant quantities of oxygen were available in the atmosphere allowing large-scale aerobic respiration. The oxygenation of the Earth occurred in several stages over some 2 billion years, the most important period of which is shown in Figure 1 of Chapter 6, but rose most rapidly after 850 million years ago. The rise of large vascular land plants is thought to have been an important cause of the rapid rise of oxygen during the Permo-Carboniferous period about 300 million years ago.

A caveat: The issues surrounding photosynthesis are not yet completely resolved. It is widely believed that the origin of anoxygenic photosynthesis predated oxygenic photosynthesis, that anoxygenic photosynthesis was more primitive than oxygenic photosynthesis, and that acquiring photosynthesis by horizontal gene transfer is more likely than losing photosynthesis. But a 2019 review paper by Cardona[17] is worth quoting,

> "It is too soon to claim that we understand how photosynthesis originated, let alone to claim that we understand the photochemistry of the earliest reaction centres to ascertain that the origin of anoxygenic photosynthesis pre-dates the origin of oxygenic photosynthesis."

The existence of large vascular land plants set the stage for the development of intelligent life. On Earth-like planets, there would have to be an analog of the primates found on Earth, where — because they descend from tree-dwellers — all primate species possess adaptations for climbing trees as well as stereoscopic vision, which allows the perception of depth and distance. In other words, this means that intelligent life would have to have the equivalent of hands; which, after they became terrestrial rather than arboreal animals, would allow them to ultimately evolve into a creature that could use fire. On Earth, as discussed earlier, the ability to use fire to

---

[17]Cardona, T., "Thinking twice about the evolution of photosynthesis", *Open Biol.* **9** (2019), 180246.

cook food is crucial to the rise of intelligence since eating raw foods can yield as little as a third of their caloric content compared to close to 100% for cooked food. On Earth-like planets there is no reason the energy requirements should differ from that on Earth.

The fact that on Earth, once adequate energy was available due to the invention of cooking, and large brains in the Homo lineage were able to evolve in only 1.5 million years, means that large brain mass and the consequent rise in intelligence are strongly selected for in an evolutionary context. On Earth, this could only occur after the rise of primates, which began some 10–15 million years after the dinosaurs became extinct. But their extinction was not inevitable and had a strong stochastic component.

Mass extinctions where more than half of extant species die off rapidly have occurred several times in the Earth's history. The largest known was the Permian-Triassic extinction some 250 million years ago. These occurrences open up new ecological niches within which surviving species could move into and rapidly evolve and diversify. In the case of the Cretaceous-tertiary extinction of some 65 million years ago it led to the rise of primates by ending the Mesozoic age of reptiles. It was primarily caused by the impact of an asteroid some 10–15 km in diameter leaving the Chicxulub crater in the Gulf of Mexico's Yucatán Peninsula perhaps coupled with the Deccan Traps volcanic eruption. This extinction event has a great deal of complexity and controversy associated with it, none of which is relevant here.

The Mesozoic age was then ended by primarily the stochastic event of a large asteroid impact. But the age of reptiles had lasted some 185 million years and could well have gone on for a much longer time without this impact. But while the timing of such events may be stochastic, the existence of them over geologic time may occur in most solar systems with a mix of large gas giant planets like Jupiter as well as smaller planets. Such a mix of different planet sizes is likely to be common.

The overall structure of solar systems may be due to the phenomena of resonance and planetary migration proposed by Fernandez,[18] who suggested that planetesimal debris in the early solar system could cause planets to migrate. This was the basis for the later work of Malhotra.[19,20]

During the migrations of the Jovian planets the Asteroid Belt was formed and the inner terrestrial planets experienced a period of enhanced meteorite impacts. Without such impacts, there would likely be an ecological stasis and in the case of Earth, the Mesozoic would likely have continued to exist until some other crisis caused ecosystem collapse. The general phenomenon of resonance and planetary migration would guarantee that in many solar systems with Earth-like planets these would also experience large meteoric impacts ultimately allowing the rise of higher-level intelligent life.

[18] Fernandez, J.A. and Ip, W.H., *Icarus* **58** (1984), 109.

[19] Lazzaro, D. *et al.* (eds.), "Solar system formation and evolution", *ASP Conf. Ser.* **149** (1998).

[20] Malhotra, R., "Resonant Kuiper belt objects: A review", *Geosci. Lett.* **6** (2019), 12. https://doi.org/10.1186/s40562-019-0142-2.

Chapter 11

Summary of the Critical Requirements for the Evolution of Intelligent Life on Earth-like Planets

For intelligent life to evolve on an Earth-like planet it must not only have a surface temperature that allows liquid water to exist, but the planet would have to be large enough for liquid water to continue to exist over billions of years.

As shown above, the most critical requirement for the evolution of high-level intelligent life is the ability to support the energy needed for a large cerebral cortex. This cannot be achieved by hunting and gathering without the ability to cook food.

For that reason, evolution on Earth-like planets must lead to the ability to use fire for cooking, which rules out planets without continents. This would in turn require the ability to manipulate objects, similar to the capability provided by hands, and an adaptation to living in a terrestrial rather than arboreal habitat.

As I have argued in Chapters 3–5, abiogenesis and the rise of eukaryotes and complex life on Earth was not a stochastic process. On Earth-like planets, except for the relatively random processes that end a period of ecological stasis, like the Mesozoic on Earth, the rise of high-level intelligence is not a stochastic process and may be far more likely than one might think.

Epilogue: Some Philosophical Thoughts on Intelligent Life on Earth-like Planets

It was only about 100 years ago that the faint nebulous objects observed in the sky were found to be other galaxies like our own when Edwin Hubble, using the 100-inch telescope on Mount Wilson, discovered Cepheid variable stars in the Andromeda Nebula. Since that time the increase in our knowledge of the universe is nothing short of spectacular! It is now known that most stars have planets, and those with Earth-like planets are all very likely to have life due to the ability of matter to form complex biomolecules.

The existence of intelligent life on Earth-like planets throughout the universe would change our fundamental understanding of the nature of reality and our existence in it. Thinking about the implications of widespread intelligent life in the universe currently falls into the domain of philosophy, which falls somewhere between theology and science. Philosophy consists of two principal components; the first is religious and ethical conceptions and the second consists of scientific considerations. Bertrand Russel put it this way, "... between theology and science there is a No Man's Land" called philosophy. He goes on to say that theology "induces a dogmatic belief that we have knowledge where in fact we have ignorance, and by doing so generates a kind impertinent insolence towards the universe".[1]

---

[1]Russel, B., *A History of Western Philosophy* (Simon and Schuster, New York 1960).

Many religions have to some extent come to terms with modern science including evolution and the origin of the universe. For example, Pope Francis has stated that, "The Big Bang theory, which is proposed today as the origin of the world, does not contradict the intervention of a divine creator, but depends on it. Evolution in nature does not conflict with the notion of creation because evolution presupposes the creation of beings that evolve." This book shows that under the appropriate conditions, life evolves naturally without requiring "the creation of beings that evolve". But the real question is: If the origin of intelligent life has a natural explanation and is widespread throughout the universe, how can one continue to believe that Earth-based humanity is central to God's creation?

The general public is not so accepting as Pope Francis of evolution or the "big bang" being the religious creation event. Nor have they had the time needed to accept the idea that humanity is not central to God's concern. To change this will involve an evolution in the minds of the collective public as to the nature of God. As a result, it will likely take humanity a long time to accept the existence of intelligent life on Earth-like planets along with its concomitant implications.

Further discussion of the relationship between science and religion and its history is given in Appendix E.

Appendix A

Bayesian and Frequentist Interpretation of Probability and Markov Processes

Statistics can be defined as the collection, organization, analysis, and interpretation of numerical data. Statistical implicitly means a stochastic process that yields random, probabilistic data. Recurrent observations can be used to define a frequency table from which a frequency curve can be drawn. This curve can be used to predict the outcome of future observations. The curve can take a variety of forms such as the binomial distribution, the normal distribution, which is an approximation to the binomial distribution, and the Poisson distribution that gives a good approximation to the binomial distribution. The latter is a discrete probability distribution that gives the probability of a given number of events occurring in a fixed interval of time, assuming that time is the variable. The frequentist interpretation of probability can then be defined as the long-run frequency of repeatable experiments. That is, probability is the limiting value of repeatable experiments. The usual example is a flipped coin where the outcome of many flips would be 50% for either heads or tails.

In a Bayesian model one can use measured data and additional information, which may not be easy to quantify, such previous studies or expert opinion. There is a relationship between Bayesian and Markov models. A Markov model is one where the future state of a system depends only on the current state of the system. Since a Bayesian interpretation can depend on previous knowledge such as previous states of a system, if this is limited to only the most recent state it becomes a Markov model. Markov models should be of particular interest in the evolution of life since Darwinian evolution works with the existing variation in the population of a species at a given time. Consequently, the discussion of Markov processes and chains given here is more extensive than Bayesian models. An introduction into how stochastic processes are used in biology has been given by Allen[1] and how statistical methods are used in molecular evolution is covered in the book edited by Nielsen.[2]

Since there are many available discussions of both Bayesian and Markov processes, only a limited heuristic presentation will be given here.

Bayesian methods use Bayes' theorem to update probabilities with new data. If A and B are two events, Bayes' theorem can be written as

$$P(A|B) = \frac{P(B|A)\,P(A)}{P(B)}, \tag{A.1}$$

where $P(A)$ is the independent probability of A occurring, $P(B)$ is the independent probability of B occurring, and $P(A|B)$ is the conditional probability of A given that B is true, and $P(B|A)$ is the conditional probability of B given that A is true. Obviously $P(B)$ cannot be zero. $P(A|B)$ is known as the posterior probability, which gives the probability of A being true after taking the new data B into account. In essence, Bays' theorem updates the prior belief $P(A)$ after introducing the new data B. Thus, Bayesian is generally used for problems where there is uncertainty or incomplete information.

A comprehensive review and primer of Bayesian statistics and modelling has been given by Rens van de Schoot *et al.*[3]

Markov Processes and Chains

Consider a system that can exist in any of a number of states, numbered 1, 2, 3, ..., j, and a stochastic process by which the system

---

[1]Allen, L.J.S., *An Introduction to Stochastic Processes with Applications to Biology* (Prentice Hall-Pearson Education, Inc. New Jersey 2003).

[2]Nielsen, R. (ed.), *Statistical Methods in Molecular Evolution* (Springer, 2005).

[3]van de Schoot, R. *et al.*, "Primer: Bayesian statistics and modelling", *Nat. Rev. Meth. Primers* **1** (2021), 1.

undergoes transitions from one state to another according to a set of transition probabilities.[4] A Markov process is a stochastic process such that the probability of entering a certain state depends only on the last state occupied. That is, the nth state transition probabilities $P_{ij}(n)$. depend only on the $(n-1)^{\text{st}}$ state.

If the Markov process goes through a finite number of steps, the result is a set of probabilities within which the system will exist in each of the j states. This set of probabilities is grouped into a row vector:

$$\pi = \{p_1, p_2, p_3 \ldots p_j\}, \tag{A.2}$$

which is referred to as a "probability vector". If the system is known to be in a particular state, the probability vector will consist of a single *1*, with the remainder of the entries being zero. After n transitions, the new probability vector will be

$$\pi_n = \{p_1^{(n)}, p_2^{(n)}, p_3^{(n)} \ldots p_j^{(n)}\}. \tag{A.3}$$

The probability vector π_n can be derived from the previous probability vector by application of a transition matrix $P(n)$

$$\pi_n = \pi_{n-1} P(n) \quad n \geq 1, \tag{A.4}$$

where the effect of multiplying the jth column of the transition matrix times the entries in row vector π_{n-1} yields the jth element of π_n. Note that the transition matrix operates on π_{n-1} from the right. The elements of the transition matrix are Pij, the probability that in one step, the system will jump from state i to state j. A new probability vector is obtained after the transition matrix operates on the initial probability vector π_{n-1}.

By repeated application of the transition matrix,

$$\pi_n = \pi_0 P(1) P(2) \cdots P(n), \tag{A.5}$$

where π_0 is the row vector corresponding to the distribution of initial probabilities.

---

[4]Marsh, G.E. and Piacesi, R., "A simplified anti-submarine warfare problem treated as a steady state Markov process", *Appl. Phys. Commun.* **8** (1988), 227–238.

A *Markov chain* is a finite, or countably infinite, Markov process where the transition matrix $P(n)$ does not depend on n.[5] Therefore, for a Markov chain $\pi_n = \pi_0 P^n$.

An example of a Markov chain is afforded by a random walk consisting of unit steps on a straight line. Let a step to the right occur with probability p, and a step to the left with probability q, except at the ends of the chain which are assumed to be absorbing. This is illustrated below for a chain with five states where the boundary states s_1 and s_5 are absorbing ($P_{11} = P_{55} = 1$). Once the system reaches s_1 or s_5 it remains there indefinitely. The state transition diagram is given in Figure A.1.

The corresponding transition matrix is

$$
\begin{array}{c c}
 & \begin{array}{ccccc} s_1 & s_2 & s_3 & s_4 & s_5 \end{array} \\
P = \begin{array}{c} s_1 \\ s_2 \\ s_3 \\ s_4 \\ s_5 \end{array} & \left(\begin{array}{ccccc}
1 & 0 & 0 & 0 & 0 \\
q & 0 & p & 0 & 0 \\
0 & q & 0 & p & 0 \\
0 & 0 & q & 0 & p \\
0 & 0 & 0 & 0 & 1
\end{array} \right)
\end{array}. \tag{A.6}
$$

Note that the sum of the entries for each row is necessarily unity.

It is possible to classify the states of a Markov chain into equivalence classes that constitute a partial ordering with respect to the possible directions in which the process can proceed. The states corresponding to these equivalence classes are either ergodic, transient, or absorbing:

- Ergodic states consist of the set of states that can be reached from any other state in the set; the set *cannot* be left once entered.

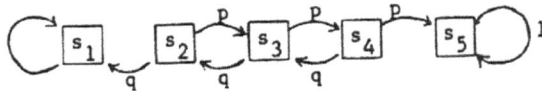

Figure A.1. Markov chain transition diagram with five states where the boundary states s_1 and s_5 are absorbing.

---

[5]Kemeny, J.G. and Snell, J.L., *Finite Markov Chains* (D. Van Nostrand Co., Inc., New Jersey, 1960).

- Transient states consist of the set of states that can be reached from any other state in the set, and which *can* be left if entered.
- Absorbing states are those which, once entered, cannot be left.

A transition matrix can be brought into canonical form where all ergodic and transient states are separately aggregated. If there are t transient states and $r - t$ ergodic states, the transition matrix takes the form

$$P = \begin{pmatrix} E & O \\ R & T \end{pmatrix}, \tag{A.7}$$

where the submatrix R relates the transient states T to the ergodic states E, and O represents a t by $r - t$ block of zero entries. In the example transition matrix above, the rearranged canonical transition matrix is shown below.

The vertical and horizontal lines divide $P_{\text{Canonical}}$ into four blocks; the upper left 2 by 2 block corresponds to E, the upper right 2 by 3 block corresponds to O, the lower left 3 by 2 block to R, and the lower right 3 by 3 block to T.

$$P_{\text{Canonical}} = \begin{array}{c} \\ s_1 \\ s_5 \\ s_2 \\ s_3 \\ s_4 \end{array} \begin{array}{cc} \begin{array}{ccccc} s_1 & s_5 & s_2 & s_3 & s_4 \end{array} \\ \left(\begin{array}{cc|ccc} 1 & 0 & 0 & 0 & 0 \\ 0 & 1 & 0 & 0 & 0 \\ q & 0 & 0 & p & 0 \\ 0 & 0 & q & 0 & p \\ 0 & p & 0 & q & 0 \end{array}\right) \end{array}. \tag{A.8}$$

Note that in this example all the ergodic states are absorbing.

A transition matrix is said to be ergodic if the powers of P approach a limit matrix L. That is,

$$\lim_{n \to \infty} P^n = L. \tag{A.9}$$

It is said to be *regular* if for some n, P^n has no zero entries. If one starts with a probability vector π, and L operates on π, the result will be a new probability vector π_∞.

$$\pi_\infty = \pi L = \lim_{n \to \infty} \pi P^n. \tag{A.10}$$

The elements of π_∞ represent the probabilities of finding the system in each state after operation with L. Let this set of probabilities be denoted by $p_1^\infty, p_2^\infty, \ldots$, so that p_i^∞ represents the fraction of time the process spends in state s_i.

L must have a structure such that when the jth column of L is multiplied by π, the result is p_j^∞. Since the sum of the elements of π must add up to unity, it follows that if all the elements in column j of the matrix L are equal to p_j^∞, the result of L operating on π will be π_∞. Thus, for a regular transition matrix P, the matrix L has the form

$$L = \begin{pmatrix} p_1^\infty, p_2^\infty & \cdots & p_n^\infty \\ \vdots & \ddots & \vdots \\ p_1^\infty, p_2^\infty & \cdots & p_n^\infty \end{pmatrix}. \qquad (A.11)$$

The definition of π_∞ implies that $\pi_\infty P = \pi_\infty$, which is an eigenvalue equation for the row vector π_∞. This row vector is often called a fixed point of P.

In practice, one desires the limiting probability distribution of a problem. There are two techniques for determining π_∞, one being solution of the eigenvalue equation, and the other being computation of a high power of P to sufficient accuracy.

Appendix B

Basics of Metabolism

It is now known that the early atmosphere of the Earth was composed primarily of carbon dioxide and nitrogen. The first organisms, arising from organic material from extraterrestrial sources or from material produced by chemosynthesis, probably used energy derived from the covalent bonds of hydrogen. Hydrogen is emitted by volcanoes and is produced when naturally occurring methane interacts with water. When hydrogen interacts with carbon dioxide it produces methane and water and releases some of the energy stored in the hydrogen.

There are a number of reactions involved in anaerobic chemosynthesis. One example is given by

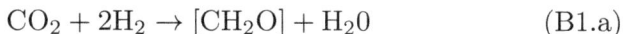

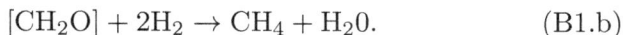

$$CO_2 + 2H_2 \rightarrow [CH_2O] + H_20 \qquad (B1.a)$$

$$[CH_2O] + 2H_2 \rightarrow CH_4 + H_20. \qquad (B1.b)$$

The first equation says that carbon dioxide and two molecules of hydrogen combine to form formaldehyde and water; the second tells us that formaldehyde can combine with two molecules of hydrogen to produce methane and water.

By adding these two equations together and cancelling the $[CH_2O]$ that appears on both sides of the resulting equation (a legitimate operation) one gets the result

$$CO_2 + 4H_2 \rightarrow CH_4 + 2H_20. \qquad (B.2)$$

This equation is known as the Sabatier reaction and is the main source of abiotic methane on the earth and other planets.

It usually requires temperatures greater than 200°C, comparable to that found in deep-sea hydrothermal vents. The reaction will occur at temperatures lower than 100°C in the presence of low concentrations of ruthenium comparable to that found in chromitite, an igneous rock — meaning rock that solidified from lava or magma in volcanic processes.

The Sabatier reaction produces enough energy to reduce a second molecule of CO_2 to organic carbon

$$2CO_2 + 6H_2 \rightarrow [CH_2O] + CH_4 + 3H_2O, \tag{B.3}$$

where Equation (B1.a), with $4H_2$ and CO_2 added to each side, and the Sabatier equation have been used. This is the main equation governing anaerobic chemosynthesis of organic material.

When the animal communities surrounding deep-sea hydrothermal vents were discovered in the late 1970s it was soon determined that the primary source of organic carbon was from symbiotic chemosynthetic bacteria. It was also soon found that such symbiotic relationships were common in shallow marine environments. And recently it was discovered that sucinid clams depend on chemosynthetic bacteria living in their gills. These clams are found in sea grass beds and supply a significant fraction of the food for Caribbean spiny lobsters. It remains to be determined how much of a role chemosynthetic primary production plays in the wider marine food web.

Another source of energy available is the reaction of hydrogen and sulfur to produce the unpleasant smelling gas hydrogen sulfide. Abundant sulfur is found near hydrothermal vents on the ocean floor. The basic reactions, which are energetically coupled in the sense that the second supplies the energy needed for the first, are

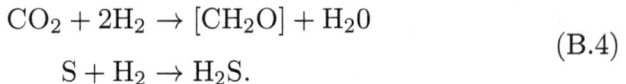

$$CO_2 + 2H_2 \rightarrow [CH_2O] + H_2O$$
$$S + H_2 \rightarrow H_2S. \tag{B.4}$$

Other theories of primitive metabolism involve the use of hydrogen sulfide to form iron pyrite FeS_2. It is argued that this could lead to a form of surface metabolism, traces of which can be found in an ancient lineage of Archaea[1] that may retain the ability to grow using the energy from pyrite formation.

Some have argued that formaldehyde is the only molecule composed of one carbon atom that is capable of generating the complex organic compounds needed for the development of life. Reactions involving formaldehyde can yield sugars and the amino acid glycine whose interaction with formaldehyde can form additional amino acid products.

The first cells would obtain energy for metabolic processes by extracting energy from a molecule by breaking it down piece by piece, a process called catabolism. This is usually achieved by use of an enzyme, and the energy can be used immediately or stored for later use. The process of building molecules to store energy is known as anabolism. Cell metabolism is a combination of the two processes.

The energy from glucose, a sugar that would have been available under prebiotic conditions, would probably been used by the first prokaryotic cells either directly or stored by forming the polymer glycogen, which is a chain of glucose molecules. The first cells also developed glycolysis[2] to store some of the energy derived from breaking down glucose. It yields two molecules of adenosine triphosphate (ATP) for each molecule of glucose. The structure of ATP is shown in Figure B.1.

The amount of energy contained in a single glucose molecule is too large for most of the chemical reactions used by living organisms, but the hydrolysis (where a molecule is broken apart by the addition of water) of ATP to ADP yields a more manageable amount of energy. The phosphate groups seen in Figure B.1 can be broken apart and the energy released used for such activities as the synthesis of more complex compounds, active transport across cell membranes, and

---

[1] Single-celled life is divided into three separate branches: archaea, prokaryotes (like bacteria) and eukaryotes (which have membrane-bound organelles like the nucleus).

[2] A sequence of ten enzyme catalyzed reactions that converts glucose into another molecule called pyruvate releasing energy to form two molecules one of which is ATP.

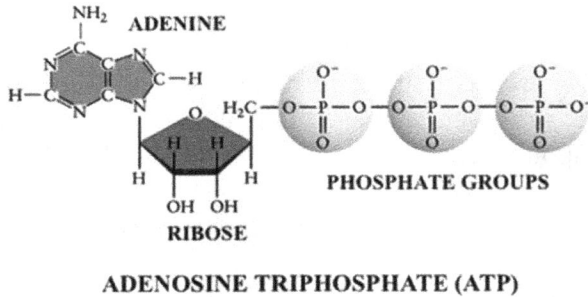

ADENOSINE TRIPHOSPHATE (ATP)

Figure B.1. Adenosine tri-phosphate. ATP stores energy in the covalent bonds between the phosphate groups and when these bonds are broken energy is released. The breaking off of one phosphate group leaves the molecule adenosine diphosphate (ADP); breaking off two groups leaves adenosine monophosphate (AMP).

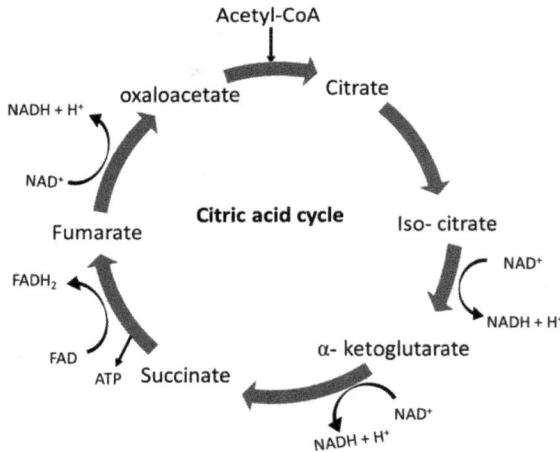

Figure B.2. The Krebs cycle, which corresponds to the equation: Acetyl-CoA + $3H_2O + 3NAD^+ + FAD + ADP + P_i \rightarrow 2CO_2 + 3NADH + 3H^+ + FADH_2 + CoA\text{-}SH + ATP + H_2O$ [Graphic after Deven Gajera's answer to a Quora question on the Krebs cycle].

muscle contraction. In modern cells, ADP is converted back to ATP by a process known as oxidative phosphorylation by mitochondria, a cellular organelle that has its own circular DNA like prokaryotes and appears to be an example of endosymbiosis.

ATP is produced using the Krebs citric acid cycle discussed in Chapter 10. A simplified diagram of the Krebs cycle is shown in Figure B.2

For an extensive description of the Krebs cycle see Berg, Tymoczko, and Stryer.[3]

---

[3]Berg, J.M., Tymoczko, J.L. and Stryer, L., *Biochemistry*, 5th edition. (W.H. Freeman and Co., New York, 2002).

Appendix C

Some Detail on Dolphin and Human Hearing

In humans, the hair cells and auditory nerve fibers from the cochlea are limited in the frequency they can respectively generate and carry in response to an audio signal, a frequency ~3 kHz being the upper limit. For frequencies above ~3 kHz, the brain relies on the tonotopically organized basilar membrane of the cochlea (see Figure C.4), which provides a spatial separation of higher frequencies. Separate nerve fibers from the different tonotopic regions of the cochlea then carry information to the (secondary) auditory cortex of the brain to form a tonotopic map where different frequencies go to adjacent regions of the cortex. While the details will certainly differ, a similar solution to the frequency limits of nerve fibers (tonotopic brain maps) must exist in dolphins.

Figure C.1 uses the auditory brain stem response to determine the low and high-frequency limit of hearing in the common dolphin.[1] The unit used to plot threshold response is the decibel, a comparison of sound intensities or energy density even though "dB re 1 μPa" appears to refer to a pressure; more precisely, "dB re 1 μPa" refers to the intensity of a plane wave of pressure equal to 1 μPa.

The common dolphin is smaller than the bottlenose dolphin and has a brain weighing about 800 g compared to 1500 g for the bottlenose dolphin. Humans have brains weighing ~1400 g.

As mentioned in Chapter 9, the very high frequencies that dolphins hear can also can be heard by humans. SCUBA divers can

---

[1]Popov, V. V. and Klishin, V.O., "EEG study of hearing in the common dolphin, *Delphinus delphis*", *Aquatic Mammals* **24(1)** (1998), 13–20.

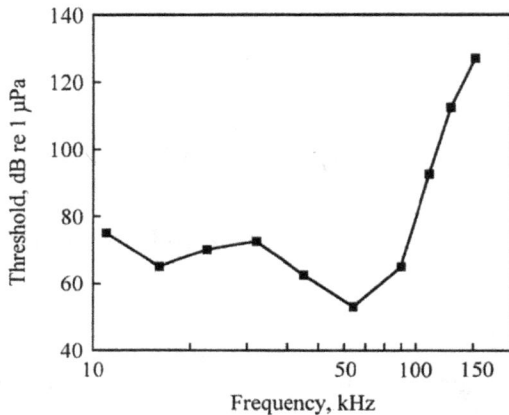

Figure C.1. Dolphin auditory brain stem threshold to tone bursts as a function of frequency. [From Popov, V.V. and Klishin, V.O., *Aquatic Mammals* **24**(1)(1998).]

hear ultrasonic frequencies to greater than 100 kHz, but have no pitch discrimination above ~20 kHz. The evolution of the dolphin auditory system has resulted in greater high-frequency hearing sensitivity and very complex auditory processing.

Individual auditory nerve fibers of the human or dolphin transfer information from only a narrow part of the audible frequency spectrum. Electrophysiological recordings of the threshold response of the nerve fibers to sound are known as tuning curves. They are plotted as the threshold intensity in dB required to achieve a response above the spontaneous firing rate of the associated neurons (in humans from essentially zero to 120 spikes/s) as a function of frequency. Tuning curves for the dolphin have been obtained by monitoring their auditory brain stem response, and are similar in shape to those seen in other mammalian species. An example of a tuning curve is shown in Figure C.2.

The method used to obtain such tuning curves is often the "masked threshold" technique: The absolute threshold of sound is the minimum detectable level of that sound when heard alone. The masked threshold is the quietest level of the signal that can be heard when combined with a masking signal. The method consists in presenting a listener with a target tone of fixed level and frequency

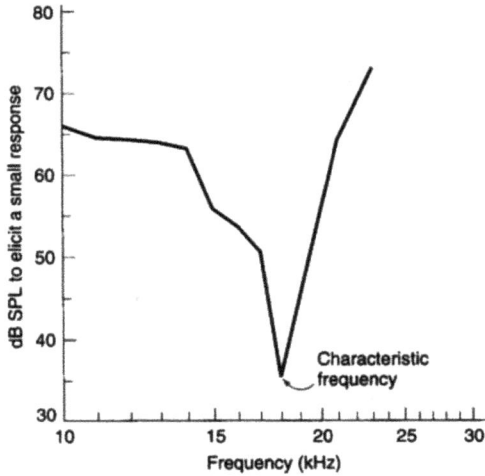

Figure C.2. An example of a tuning curve. SPL means Sound Pressure Level and the terminology dB SPL is generally defined for measures of human hearing as $20 \log_{10} p_1/p_0$ where p_0 is the reference value in μPa (figure from Wikimedia open). The values on the ordinate are often inverted and shown in negative values of dB. It then means that p_0 is greater than p_1.

and measuring the power that a second tone must have to mask the target as a function of the frequency of the masking tone. An example of tuning curves for the common dolphin is shown in Figure C.3.

The "quality" of tuning curves is often given in terms or Q_{10}, the center frequency divided by the bandwidth at a level of 10 dB above the bottom tip of the curve. With few exceptions, the quality is almost constant across a wide frequency range so that in humans an octave occupies a constant interval ~ 4 mm along the basilar membrane of the cochlea. Thus, in mammals in general, the auditory system can be thought of as a set of frequency tuned bandpass filters where Q_{10} is essentially constant across the auditory frequency range. Known exceptions include one species of bat and the small porpoises classed as *Phocoenidae*.[2] The Q_{10} value for the bottlenose dolphin is

---

[2]Popov, V.V. *et al.*, "Nonconstant quality of auditory filters in the porpoises, *Phocoena phocoena* and *Neophocaena phocaenoides* (Cetacea, Phocoenidae)", *J. Acoust. Soc. Am.* **119**(5) (2006).

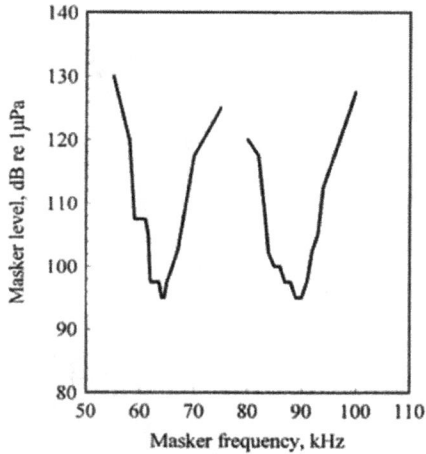

Figure C.3. Tuning curves for the common dolphin at the probe frequencies of 64 kHz and 90 kHz. [From V.V. Popov and V.O. Klishin, "EEG study of hearing in the common dolphin, *Delphinus delphis*", *Aquatic Mammals* **24**(1) (1998), 13–20.]

about twice as large as that for the common dolphin. Humans have a Q_{10} value close to that of the common dolphin.

Tuning curves are associated with different regions of the basilar membrane of the cochlea as shown in Figure C.4. Although the figure is for the human cochlea, a similar layout would exist in the dolphin cochlea albeit at higher frequencies.

The vestibulocochlear nerve, known as the eighth cranial nerve, transmits sound and equilibrium information from the inner ear to the brain. The dolphin auditory nerve has several times as many fibers as the human eighth nerve, and the fiber diameters are also about twice as large as in humans, which about doubles their speed of signal propagation. The auditory tonotopic map in the dolphin brain has been displaced from the temporal to the parietal lobe (above the temporal lobe and behind the frontal lobe) and dorsal part of the hemisphere.[3]

---

[3]Ridgway, S.H., "The auditory central nervous system of dolphins", in *Hearing by Whales and Dolphins*, eds. Au, W.W.L., Popper, A.N. and Fay, R.R. (Springer-Science+Business Media, LLC, 2000).

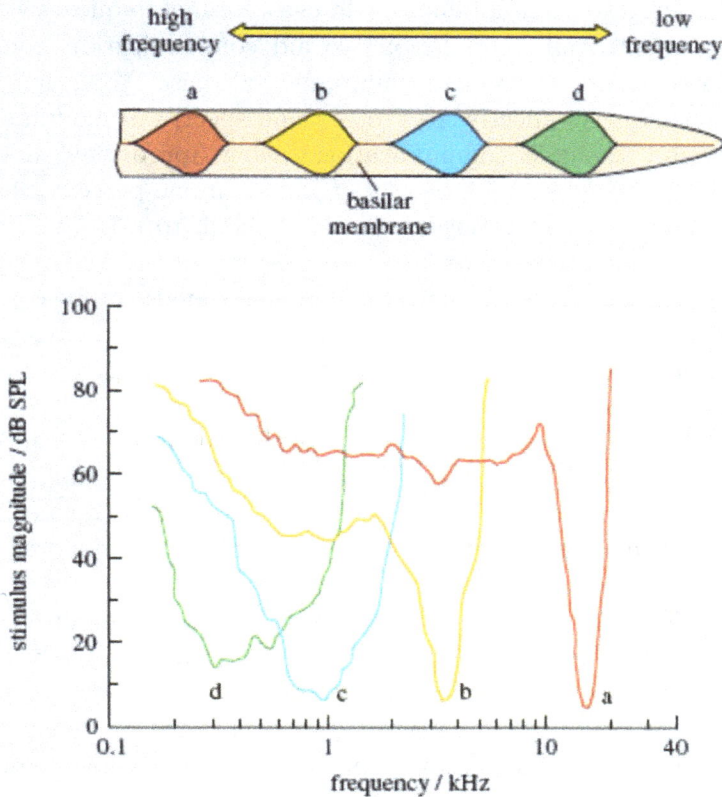

Figure C.4. Color-coded tuning curves and associated areas of the basilar membrane of the human cochlea. The basilar membrane of the cochlea is shown unrolled. [Figure from Wikimedia open.]

Dolphin sound production and reception are directional and production can be both amplitude and frequency modulated. The pointing of the sound is achieved by reshaping the dolphin sound focusing organ (the melon), and frequency, intensity, and duration can all be modulated independently. Since, as discussed above, dolphin tuning curves can be modeled as a series of overlapping bandpass filters that vary in sensitivity and bandpass region. They would be distributed across the range of dolphin hearing.

This suggests that dolphins could use a form of frequency hopping or spread-spectrum modulation[4] to aid communication fidelity in a wide spectrum of ambient environmental noise. Yang and Yang[5] have shown the effectiveness of spread-spectrum modulation in underwater acoustic communication, using appropriate detection methods, to have a bit error rate of less than one percent when the input signal-to-noise ratio is as low as -11 db to -14 db. The "bit error rate" measure comes from the fact that most communication is digital in nature. Since digital and analog signals can be converted into each other, Yang and Yang's measurements would be relevant for dolphin communication, should they use a form of spread-spectrum communication.

In humans, as pointed out by Vanthornhout *et al.*[6] and R.V. Shannon *et al.*,[7] the primary factor in speech intelligibility is the temporal envelope having a modulation frequency below 20 Hz. The most important frequency range is 4–8 Hz corresponding to the average frequency of speech.

The history of cochlear implants for hearing impaired people illustrates the importance of the temporal envelope. Early models of such implants had only one channel so that they could only provide a single time varying waveform. Nonetheless, recipients of such implants could understand speech; frequency is not the primary factor in speech comprehension.

---

[4]Spread spectrum means the signal bandwidth is much wider than the message bandwidth while frequency hopping rapidly switches the carrier carrying the modulation between different frequency channels. The most widely used spread-spectrum methods are pseudonoise modulation, frequency hopping, or a hybrid of these two methods. The spread-spectrum technology is now widely used for secure communications. Frequency hopping was invented by the actress Hedy Lamarr in 1941. Her patent for radio frequency hopping was granted on 10 June 1941. Her WW-II contribution was only publicly acknowledged decades later.

[5]Yang, T.C. and Yang, W-B., "Low probability detection underwater communication using direct-sequence spread spectrum", *J. Acoust. Soc. Am.* **124** (2008), 3632–3647.

[6]Vanthornhout, J. *et al.*, "Speech Intelligibility predicted from entrainment of the speech envelope", *J. Assoc. Res. Otolaryngol* (2018), 181–191.

[7]Shannon, R.V. *et al.*, "Speech recognition with primarily temporal Cues", *Science* **270** (1995), 303–304.

Figure C.5 shows the upper and lower envelopes of an example of human speech. Note the near mirror symmetry of the two signals. The Fourier transform used to produce the spectrum shown in Figure C.6 combines the two signals. These figures were produced during the course of a hearing experiment that used electric currents corresponding to the speech envelopes to induce cortical entrainment intended to enhance speech comprehension. The work was stimulated by the article by Wilsch[8] where non-invasive transcranial alternating current stimulation (tACS) was used in an attempt to increase speech comprehension. Both the Wilsch experiment and mine were unable to significantly increase speech comprehension, or a put by her, "... the results do not inform us about the actual benefit of envelope-tACS for speech comprehension".

The auditory cortex of non-human primates is composed of areas known as the core, belt, and parabelt regions. In humans, the primary auditory cortex corresponds to the core regions. Kubanek *et al.*[9]

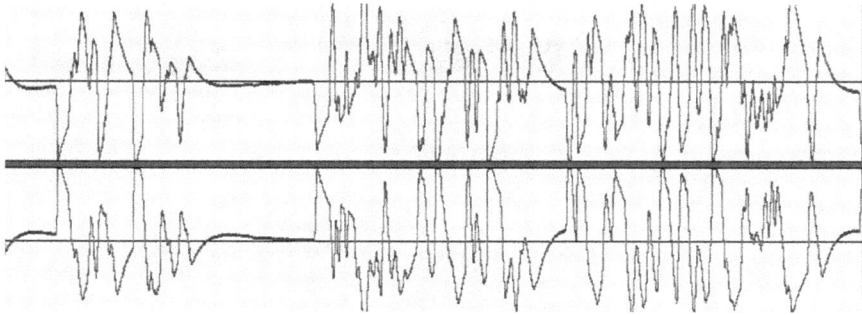

Figure C.5. The upper and lower temporal envelopes of a sample of human speech. The original auditory signals were filtered with a second-order lowpass Butterworth filter with a corner frequency $\sim$13 Hz in order to obtain the envelopes, which should be noted are not quite symmetric. [G.E. Marsh, unpublished.]

---

[8]Wilsch *et al.*, "Transcranial alternating current stimulation with speech envelopes modulates speech comprehension", *J. NeuroImage* **172** (2018), 766–774

[9]Kuganek J. *et al.*, "The tracking of speech envelope in the human cortex", *PLOS One* **8**, e53398.

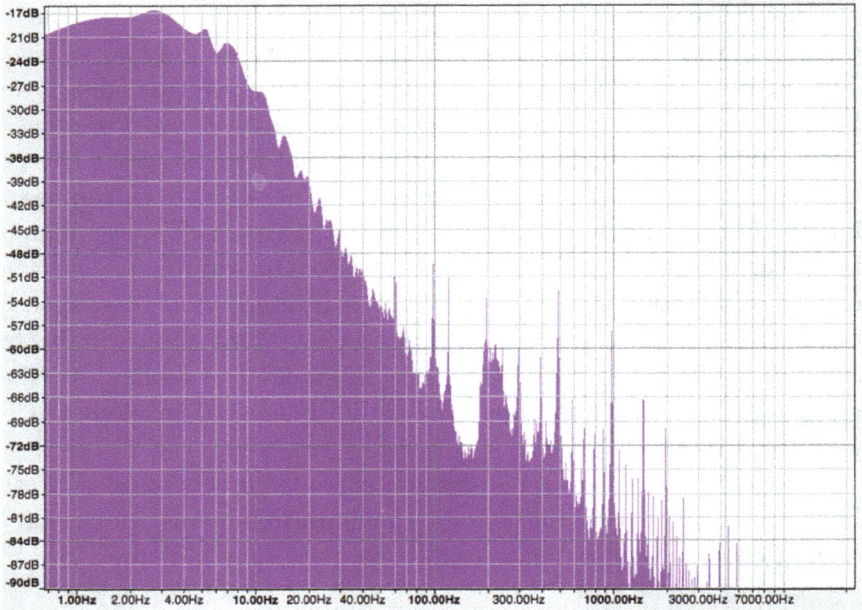

Figure C.6. The spectrum of the combined envelopes is shown in Fig. 5. [G.E. Marsh, unpublished.]

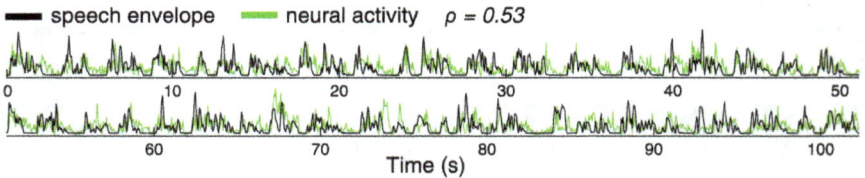

Figure C.7. Gamma wave tracking in the human auditory cortex. The neural signal has been scaled to the magnitude of the envelope. ρ is the Spearman correlation coefficient. Gamma waves in the brain have a frequency range of about 30–80 Hz, while high-gamma waves have the range of 80–200 Hz.

found that the human non-primary auditory cortex faithfully tracks the speech envelope, a phenomenon often called cortical entrainment or phase-locking. This is shown in Figure C.7.

Experiments with bottlenose dolphins show that 14 ms tone bursts, amplitude modulated at 600 Hz, evoked a strong auditory neural response that is phase-locked with the envelope of the sound. The modulation index used in the experiment was unity (100% modulation), meaning that the amplitude of the carrier wave was the same as that of the modulating wave. The neural response was monitored as is done when recording an EEG: A recording suction cup electrode was placed on the dolphin's skin about 6 cm behind the blowhole with a ground electrode behind that followed by a reference electrode.[10] That is, the "recording electrode" and the "reference electrode" go to the inputs of a differential amplifier so that in-phase inputs cancel and out-of-phase inputs are amplified with respect to the grounded electrode. This allows signal amplitudes in the microvolt range to be recorded.

Figure C.8 shows the auditory brain stem response to a "click".[11]

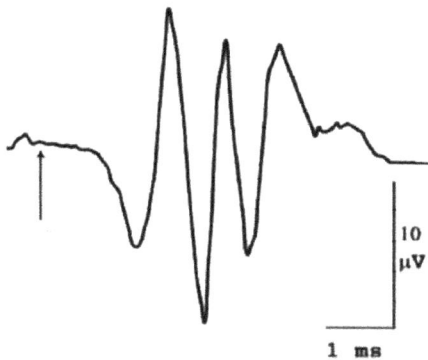

Figure C.8. Auditory brain stem response to a 120 dB re 1 μPa click produced by a piezoceramic transducer activated by 5 μs long rectangular pulses (the arrow corresponds to when the signal reached the dolphin's head where the response signal was recorded). The latency can be seen to be 0.75 ms. Recordings here and in Fig. 9 were from a common male dolphin 1.54 m long (an adult reaches 1.9-2.5 m) that was ill and died after four days of monitoring. [Adapted from Fig. 1 of Ref. 116.]

---

[10]Cook, M. L. H., "Behavioral and auditory evoked potential (AEP) hearing measurements in odontocete cetaceans", Graduate Theses and Dissertations, 2006. https://scholarcommons.usf.edu/etd/2489.

[11]Popov, V.V. and Klishin, V.O., "EEG study of hearing in the common dolphin, *Delphinus delphis*", *Aquatic Mammals* **24**(1) 1998, 13–20.

A 0.75 ms latency period corresponds to the ability of the auditory nerve fibers to carry signals up to at least 1300 Hz. The paper by Popov and Klishin contains another figure that is very important for interpreting dolphin signals.

When measuring the response to paired clicks having a separation of 2 ms (corresponding to 500 Hz if part of a continuous series) the responses just merge, as can be seen in Figure C.9. Thus, 300 Hz is likely to be the maximum frequency for communication purposes. There is a peak around this frequency in the spectra shown in Figure C.9.

As mentioned above, dolphins also have an envelope following response as can be seen in Figure C.10. The fact that the modulation following response appears up to about 1250 Hz in dolphins does not imply that the maximum frequency used for communication is greater than the $\sim$300 Hz implied by the measurements shown in Figure C.9.

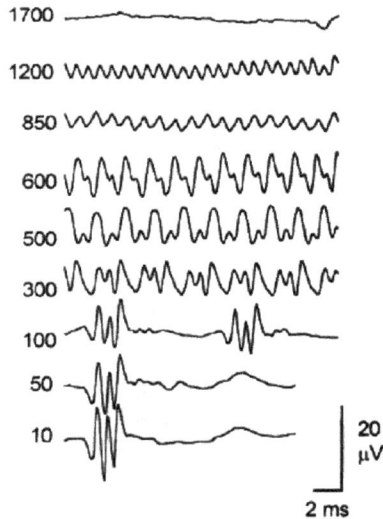

Figure C.9. Auditory brain stem response to rhythmic clicks of different rates (indicated by number per second to the left of the traces). The sound intensity was 120 dB re 1 μPa. Note that the response smooths above 200 s^{-1} (or 200 Hz) to form a sinusoidal response at 300 Hz. [Figure 5 of Ref.116.]

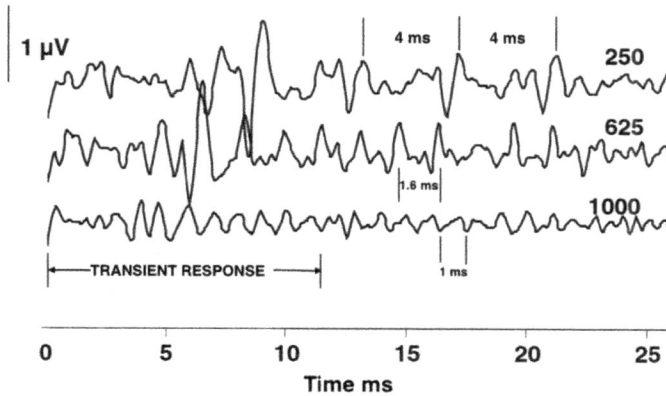

Figure C.10. The rate following response of a white-beaked dolphin to click stimuli. The numbers on the right-hand side correspond to the modulation frequency in Hz; the click time interval corresponding to 250 Hz is 1 click every 4 ms, 625 Hz to 1.6 ms, and 1 kHz to 1 ms. Responses are an average of 1000 recordings. The entrainment begins after a transient response. [Adapted from Mooney, T.A. *et al.*,*J. Comp. Physiol. A* **195** (2009), 375–384.]

A Fourier analysis of the 250 Hz and 625 Hz response traces shown in Figure C.10 clearly shows harmonics up to ∼1250 Hz.

Dolphins can detect audio signals with a very fine time resolution, their integration times for a single short signal being on the order of 200–300 μs. Integration time is a measure of the ability to isolate individual acoustic events. A longer integration time corresponds to a reduction in temporal resolution. This fine time resolution allows dolphins to use high-frequency broadband clicks of microsecond duration to visualize objects around them in terms of distance, shape, orientation, and composition. This is because the echo signal is amplitude modulated and several milliseconds in duration containing a great deal of non-spectral information. With reference to Figure C.10, note that 200–300 μs taken as a frequency corresponds to 500–330 Hz.

The ocean is a complex medium where temperature, density, sound velocity, and salinity vary with depth. There are many factors that affect the return echo from a dolphin's clicks. Sound does not travel in a straight line in the ocean. For example, the mixed

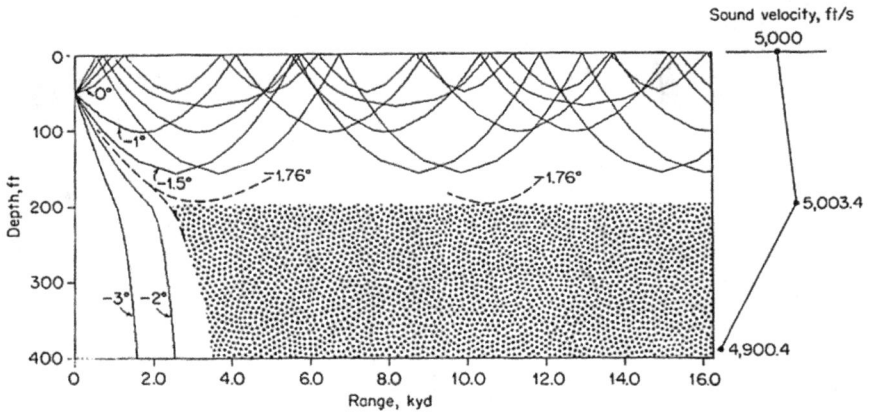

Figure C.11. Ray diagram for the transmission of sound from a 50-ft source in a 200-ft mixed layer. [From Urick, R.J., *Principles of Underwater Sound*, 3rd edition. (McGraw-Hill Book Co., New York, 1983), Fig. 6.1 (with minor corrections)]

isothermal layer, which extends to about 60 m, forms a sound channel where sound emitted at a shallow angle returns to the surface where it is again reflected — see Figure C.11; the return signal from moving object would also be doppler shifted. Over short distances the curved ray paths would not matter, but for communication over longer distances it would, for example for a female dolphin locating a lost calf who is sending a distress signal.

Multipath, where there is more than one propagation path for sound between the dolphin and the sound scattering object, is an important factor in spreading out the return signal. In shallow water, the dolphin must also take into account scattering from the seabed.

Appendix D

Introduction to Symbiosis

In the body of this book, symbiosis appears mostly in the context of endosymbiosis in Chapter 4 on the prokaryote–eukaryote divide. But symbiosis continues to play a major role after the development of eukaryotes up to this day.

As a reminder, here is a précis of endosymbiosis: It is the uptake of a prokaryote by another living cell. At first there was the genetic integration of a non-photosynthetic alphaproteobacterium, which was followed later by a photosynthetic cyanobacterium into a eukaryotic cell. This was one of the most important events in the evolution of complex life.

Symbiosis in general is usually defined as the relationship between two different organisms in a close physical association to the advantage of both.[1] In fact, this is far too narrow. Symbiosis between species is in reality a fundamental feature of evolution. As put by Gilbert, Sapp, and Tauber, "What could 'individual selection' mean if all organisms were chimeric, and there were no real monogenetic individuals?"[2] In humans, thousands of bacterial species have an intimate association with our eukaryotic cells so that some 90% of the cells in our bodies are bacterial.

Symbiosis came to be known through lichens, which grow over some 8% of our planet's surface. Lichens, and their association with

---

[1]his type of symbiosis is known as mutualism. There are two other accepted types: parasitism and commensalism, where one species is helped and the other is neither positively or negatively affected.

[2]Gilbert, S.F., Sapp, J., and Tauber, A.I., "Asymbiotic view of life: We have never been individuals", *Quart. Rev. Biol.* **87** (2012), 325.

rock, are important part of the process known as weathering, where they help break up rock by their growth and then dissolve and digest it. Silicate weathering is part of the Earth's long-term carbon dioxide regulation. It appears, however, that weathering can also release carbon dioxide as rock organic carbon, previously sequestered from the atmosphere by photosynthesis, can also be released by oxidation near the surface of the Earth.[3]

Symbiotic relationships range from the mitochondria in host eukaryotic cells to looser relationships between separate organisms. Marine invertebrates have symbiotic algae and animals have their gastrointestinal tract microbial symbionts. And a more exotic example are the tube worms found at hydrothermal vent sites on mid-ocean ridges. Adult tube worms obtain all their energy and food from intracellular symbiotic bacteria. They have a unique hemoglobin in their blood that brings oxygen and sulfide to the bacteria from seawater. The bacteria synthesize carbon compounds that are the main source of nutrition for the worms. Some additional examples and details have been given by Dimijian.[4]

It was mentioned in Chapter 10 that only when fungi and algae began forming symbiotic mycorrhizal relationships did life begin to flourish on the continents. There are seven types of mycorrhizal fungi and plants need to have these fungi colonize their roots. The hyphae of the fungus greatly increase the surface area available for nutrient and water absorption thereby maximizing the plant's access to essential compounds and elements. In return, they get nutrients. These symbiotic relationships have evolved into different forms and now encompass most land plants and fungal groups.

While the symbiotic relationship that forms lichens creates a form unlike that of the individual members of the symbiosis, mycorrhizal relationships do not. The members are clearly recognizable. In this type of symbiosis, a single plant may have mycorrhizal relationships

---

[3]Zondervan, J.R. *et al.*, "Rock organic carbon oxidation CO_2 release offsets silicate weathering sink", *Nature* **623** (2023), 329.

[4]Dimijian, G.G., "Evolving together: The biology of symbiosis, part 1", *BUMC Proc.* **13** (2000), 217–226.

with many fungi at the same time. They provide plants with some 80% of their nitrogen, almost all of needed phosphorous, as well as other nutrients, while plants in return allocate as much as 30% of the carbon they produce from photosynthesis.

There are many modern examples of symbiosis, perhaps one of the best known is between coral and algae. Reef building corals are marine animals whose colors come from zooxanthellae[5] algae. The zooxanthellae provide nutrient sugars for the coral with oxygen as a byproduct, while the zooxanthellae obtain essential nutrients and shelter from the coral. Coral bleaching occurs when stress causes the coral to expel their algae. However, living without algae for too long can be fatal.

Another famous example is a type of what might be called a social or behavioral symbiosis that occurs between birds known as honeyguides and humans. In Tanzania the Hadza people work with honeyguides to guide people to bee nests. The honeyguides notify people when they have found a bee nest with special call, which the humans answer with calls passed down from one generation to the next. The birds learn the distinct whistles and calls of the human foragers that live near them. Once arriving at the nest, the humans will chase the bees away and break the nest up to obtain the honey; afterwards, the birds eat the beeswax, eggs and larvae left behind. In Tanzania and Mozambique honeyguides respond preferentially to their local honey-hunting partners than to the calls of honey hunters from other regions.[6]

---

[5]From Wikipedia: Zooxanthellae is a colloquial term for single-celled dinoflagellates that are able to live in symbiosis with diverse marine invertebrates including demosponges, corals, jellyfish, and nudibranchs.

[6]Spottiswoode, C.N. and Wood, B.M., "Cultjurally determined interspecies communication between humans and honeyguides", *Science* **382** (2023), 1155–1158.

Appendix E

Discussion of the Relationship Between Science and Religion[*]

The epilogue to this book pointed out that if intelligent life exists on Earth-like planets, it brings up many issues concerned with the relationship between science and religion. This appendix is intended to explain this relationship in its historical context.

It was only a few hundred years ago that the Enlightenment allowed humanity to understand the world in scientific rather than religious terms, and the Industrial Revolution, with its enormous impact on productivity, permitted a large fraction of the population, at least in the developed world, to engage in something other than production, storage, and distribution of food.

One of the most important contributions of the Enlightenment to the future development of modern society was made by Francis Bacon in the 17th century. His ideas changed the very relationship between humanity and nature: he introduced the concept of empiricism and popularized the inductive method of scientific inquiry. This, of course, is the basis of the scientific method, an approach to nature that was unheard of in his time. In Bacon's words: "At the foundation we are not to imagine or suppose, but to *discover* what nature does or may be made to do." Loren Eiseley,[1,2] who has written extensively

---

[*]This appendix has been abstracted with minor changes from my article "Weinberg's Lament: Science and Religion", *Sci. Religion Cult.* **3** (2016), 49–54.

[1]Eiseley, L., *The Invisible Pyramid* (Charles Scribner's Sons, New York, 1970), p. 68.

[2]Eiseley, L., *The Man Who Saw Through Time* (Charles Scribner's Sons, New York, 1973).

on Francis Bacon, describes Bacon as "preeminently the spokesman of *anticipatory* man. The long reign of the custom-bound scholastics was at an end. Anticipatory analytical man, enraptured by novelty, was about to walk an increasingly dangerous pathway".

This "dangerous pathway" has led to a strong reaction against the Enlightenment. As put by Isaiah Berlin[3] in his essay *The Counter-Enlightenment,*

> "The proclamation of the autonomy of reason and the methods of the natural sciences, based on observation as the sole reliable method of knowledge, and the consequent rejection of the authority of revelation, sacred writings and their accepted interpreters, tradition, prescription, and every form of non-rational and transcendent source of knowledge, was naturally opposed by the Churches and religious thinkers of many persuasions."

This is the branch of the Enlightenment whose impact on society was ultimately to liberate most people in the western world from the terrible fear generated by rampant superstition, but one should remember that the Enlightenment itself evolved from the anti-scholastic Platonists of the Renaissance, which gave us so much great art and music and other elements of culture.

The Enlightenment also led to Darwinian evolution and a perceived conflict with religion: If the origin of life, and humanity in particular, has a natural explanation, how can one believe in the immortal soul, or that humanity is central to God's creation? As put by Omar Khayyám — a doubter of long ago — in two of the quatrains of his *Rubáiyát*:

> "There was a door to which I found no key:
> There was a veil past which I could not see:
> Some little talk awhile of me and thee
> There seem'd — and then no more of thee and me.
>
> Then to the rolling heav'n itself I cried,
> Asking, "What lamp had destiny to guide

---

[3]Berlin, I., *The Proper Study of Mankind* (Farrar, Straus and Giroux, New York, 1989), p. 243.

Her little children stumbling in the dark?"
And — "A blind understanding!" heav'n replied."

Khayyám's "blind understanding" is surely in the realm of faith, which in turn leaves open the possibility of revelation. Revelation (in at least Islam, Christianity and Judaism) with its eternal truths is incompatible with science, which requires reproducibility. But there is a form of revelation — not based on theophany — that *is* compatible with science.

As put by James Carroll[4] in his brilliant history, *Constantine's Sword*, "the truth of our beliefs is revealed in history, within the contours of the mundane, and not through cosmic interruptions in the flow of time. Revelation comes to us gradually, according to the methods of human knowing. And so revelation comes to us ambiguously. Certitude and clarity are achieved only in hindsight, and even then provisionally." Since it is this provisional nature of knowledge that is also the essence of scientific knowledge, religious people who find themselves able to accept Carroll's characterization of revelation should have no difficulty accepting the findings of modern science — those findings reflect the will of God. It is worth noting that Carroll was a Catholic priest before taking up writing as a career.

Carroll's characterization of revelation makes it clear that in his view God does not exist in the sense of western thought; that is, there is not *per se* a "revealer". His characterization is perhaps closest to that of Spinoza; as put by Bertrand Russell,[5]

"Individual souls and separate pieces of matter are, for Spinoza, adjectival; they are not *things*, but merely aspects of the divine Being. There can be no such personal immortality as Christians believe in, but only that impersonal sort that consists in becoming more and more one with God. Finite things are defined by their boundaries, physical or logical, that is to say, by what they are *not*: 'all determination is negation.' There can be only one Being who is

---

[4]Carroll, J., *Constantine's Sword: The Church and the Jews* (Houghton Mifflin Co., Boston, 2001), p. 172.
[5]Russell, B., *A History of Western Philosophy* (Simon and Schuster, New York, 1960), p. 571.

wholly positive, and He must be absolutely infinite. Hence Spinoza
is led to a complete and undiluted pantheism."

Here, pantheism should be interpreted as the doctrine of identifying God with the various forces and workings of nature.

The "blind understanding" of Khayyám is not enough for most people to bridge the gap between revelation and scientific discovery, and not even for some scientists. Science has been able to reveal the evolution of the universe back to the first moment of its coming into existence, but cannot offer any explanation for what Fred Hoyle derogatorily called the "big bang," other than it might have been a random and meaningless quantum fluctuation. What this "fluctuation" was supposed to have taken place in, since neither space nor time, as we understand it, had yet come into existence, is left unanswered. The problem is that the universe's coming into existence is a *sui generis* event, which places it outside the domain of the scientific method.

I said earlier that the Enlightenment had an impact on society that was ultimately to liberate most people in the western world from the terrible fear generated by rampant superstition. This is true in the sense that it led to modern science, which transformed the western world in a few hundred years; nothing comparable has occurred in human history. But the Enlightenment has also been extended to other branches of knowledge and misinterpreted to mean that there are eternal, timeless truths that implicitly govern moral, economic, political and the social spheres of human activity. All such theories are contradictory to the fundamental precepts of science.

The concept of timeless truths has a long intellectual history. Plato strongly emphasized timeless truths and Aristotle in the *Nicomachean Ethics* maintained that one of the highest virtues was the contemplation of such timeless truths. Pre-Enlightenment religious thought was also based on eternal truths, as were more modern social theories. Hegel believed that objective concepts and principles that govern human society exist, and that history evolves as a dialectical process. Marx identified these principles as material relations between classes, which were governed by general laws.

These moral and political constructs based on scientific theories of economics, sociology and psychology have failed abysmally in the 20th century causing far too much suffering and many humanitarian crises.

History has shown that the concept of empiricism and the inductive method of scientific inquiry have only limited applications in other areas of human endeavor. There is no morality implicit in science, and the methods of science have led to much quantification but few advances in the understanding of the economic, political and social aspects of human existence.

That science cannot provide a moral framework does not mean that scientists do not have a ethical responsibility to clearly inform the general public about the implications for society of their discoveries. This was true in the 20th century, for example, with regard to the discovery of nuclear fission and fusion and is especially true in the rest of the 21st century with the ongoing revolution in biology and the growing ability to modify existing organisms and create new ones.

If the sole reliable method of gaining knowledge is through the autonomy of reason and the methods of the natural sciences, where does that leave us? It leaves us with our ignorance about the meaning behind the existence of the universe. Faith and tradition can offer solace, but the validity of any "truths" offered by faith cannot be proven by science or reason alone.

Appendix F

Emergent Behavior and the Biological Sciences

Emergent behavior that appears at a given level of organization may be characterized as arising from an organizationally lower level in such a way that it transcends a mere increase in the behavioral degree of complexity. It is therefore to be distinguished from systems exhibiting chaotic behavior, for example, which are deterministic but unpredictable because of an exponential dependence on initial conditions. In emergent phenomena, higher-levels of organization are not determined by lower levels of organization; or, more colloquially, emergent behavior is often said to be "greater than the sum of the parts".

Complex systems, and in particular biological systems, often display emergent behavior. Associated with this phenomenon is a sense of the mysterious: the emergent properties of the collective whole do not in any transparent way derive from the underlying rules governing the interaction of the system's components. Unfortunately, there is not even a universally acknowledged definition of emergence. Nor do the concept and its explication in the literature constitute an organized, rigorous theory. Instead, it is more of a collection of ideas that have in common the notion that complex behavior can arise from the underlying simple rules of interaction. There are, however, some very interesting attempts at definition and theory. See, for example,

Ryan[1] and references cited therein, and the book edited by Clayton and Davies.[2]

The philosophical literature[3] separates emergence into two basic categories: *strong emergence*, where higher-level processes cannot *in principle* be derived from lower-level processes; and *weak emergence*, where there is no known way to derive higher-level processes from lower-level processes, but it has not been shown that this cannot be done. It should be noted, as should become clear in what follows, that an acceptance of the concept of emergence does not imply that a denial of reductionism must follow. It does however mean we must be careful about what we mean by reductionism.

Ernst Mayr[4] in his monumental work *The Growth of Biological Thought* characterizes emergence as follows:

> "Systems almost always have the peculiarity that the charac-
> teristics of the whole cannot (not even in theory) be deduced
> from the most complete knowledge of' the components, taken
> separately or in other partial combinations. This appearance of
> new characteristics in wholes has been designated as *emergence*.
> Emergence has often been invoked in attempts to explain such
> difficult phenomena as life, mind, and consciousness. Actually,
> emergence is equally characteristic of inorganic systems. As far
> back as 1868, T. H. Huxley asserted that the peculiar properties of
> water, its 'aquosity,' could not be deduced from our understanding
> of the properties of' hydrogen and oxygen. The person, however,
> who was more responsible than anyone else for the recognition
> of the importance of emergence was Lloyd Morgan. There is no
> question, he said, 'that at various grades of organization, material
> configurations display new and unexpected phenomena and that

---

[1]Ryan, A., "Emergence is coupled to scope, not level", (2006), arXiv:nlin/0609011 v1.

[2]Clayton, P. and Davies, P., *The Re-Emergence of Emergence* (Oxford University Press, Oxford, 2006).

[3]Antonella, C. and O'Connor, T. (eds.), *Emergence in Science and Philosophy*, Routledge Studies in the Philosophy of Science (Routledge, New York, 2010); Mark A. B. and Humphreys, P. (eds.), *Emergence*, Contemporary Readings in Philosophy and Science (MIT Press, Cambridge, 2008).

[4]Mayr, E., *The Growth of Biological Thought: Diversity, Evolution, and Inheritance* (Harvard University Press, Cambridge, Massachusetts, 1982).

these include the most striking features of adaptive machinery.'
Such emergence is quite universal and, as Popper said, 'We live
in a universe of' emergent novelty'. Emergence is a descriptive
notion which, particularly in more complex systems, seems to resist
analysis. Simply to say, as has been [done],[5] that emergence is due
to complexity is, of course, not an explanation. Perhaps the two
most interesting characteristics of new wholes are (1) that they,
in turn, can become parts of still higher-level systems, and (2)
that wholes can affect properties of components at lower levels.
The latter phenomenon is sometimes referred to as 'downward
causation'. Emergentism is a thoroughly materialistic philosophy.
Those who deny it, like Rensch, are forced to adopt pan-psychic or
hylozoic theories of matter." [references deleted]

While the concept of emergent behavior may be difficult to define,
it is easy to intuitively understand. One defines a set of rules for
the interaction of a class of objects with each other and with the
environment, and within the constraints set by the rules, the resulting
observed behavior is often very complex, transcending the simplicity
of the rules themselves. A concrete and practical example is the class
of artificial insects developed at MIT,[6,7] particularly the emergent
behavior displayed when central control modules are eliminated and
the robots are allowed to "learn" their behavior.

Randall Beer, Hillel Chiel, and Leon Sterling have done similar
work on simulating insect-like neural networks on a digital computer
at Case Western Reserve University.[8] The body plan they used
is based on the American cockroach and the artificial insect they
created walks with changes in gait that are an emergent property
similar to the changes in gait made by real insects as they change
their speed. That is, these gait changes are not "wired into" the
circuitry. This complex behavior is achieved with only six "neurons"
per leg, and two command neurons that are wired to all six legs.
A six-legged robot duplicating the neural circuitry used in the

---

[5]Original has the word "clone".

[6]Dewdney, A.K., "Insectoids invade a field of robots" *Sci. Am.* (July 1991).

[7]Waldrop, M.M., "Fast, cheap, and out of control", *Science* **248** (1990), 959–961.

[8]Beer, R.D., Chiel, H.J. and Sterling, L.S., "An artificial insect", *Am. Sci.* **79**
(1991), 444–452.

computer simulation was constructed and it displayed a range of gaits similar to those of the simulated insect.

Deborah Gordon has given another example of emergence[9,10] in her studies of the complex behavioral patterns exhibited by ant colonies. These behaviors, and the changes in behavior as the colony grows and ages, seems to be based on the decisions of individual ants that operate with a relatively simple set of rules based on social contact and the environment within which the individual finds itself.

The various tasks carried out in an ant colony are accomplished without direction, without the chain of command implicit in an hierarchical organization. A single set of rules, at the level of the individual, is responsible for the behavior of the colony as a whole. As an ant colony ages, individual responses are refined to adapt to changes in the colony's environment. Such refinement represents a type of social learning, with the memory trace being stored in a holographic fashion across the individuals of the colony. Small modifications of the rules of interaction can lead to significant changes in the behavior of the colony as a whole. Of course, this discussion is oversimplified in that real ant colonies are far more complex and the inhabitants of ant colonies have specialties so that there may be several subsets of rules rather than a single overarching set. There exist many other examples of emergence.

The emergent behavior of a system, while it is determined by the elements of the system and the rules of interaction between them — and perhaps with the environment, is not contained explicitly in any of the rules or elements themselves, nor is the behavior explained by a simple summation over the components making up the system. Emergent behavior is characterized by being "greater than the sum of the parts."

While emergent behavior is completely dependent on the set of rules governing the interaction between the elements of a system,

---

[9]Gordon, D.M., "The development of organization in an ant colony", *Am. Sci.* **83** (1995), 50–57.

[10]Gordon, D.M., *How an Insect Society is Organized* (W.W. Norton & Co., New York, 1999).

a key question is whether, at the level of the emergent behavior, new rules of interaction appear that are not, in a fundamental sense, predictable. Very simple mathematical models can exhibit extremely complex dynamics even though the behavior is completely deterministic. They have been successfully used to model the dynamics of systems in a variety of fields, including electronics, mechanics, and biology.[11] In each of these diverse areas, even when the behavior is chaotic (where the dynamics depends exponentially on changes in initial conditions), and the dynamical trajectories may look like random noise, the behavior is deterministic. In terms of emergence, one might characterize such chaotic, deterministic behavior as perhaps fitting into the category of weak emergence.

Mathematics can also provide examples of truly emergent properties. A Möbius strip, for example, is a one-sided non-orientable surface with one edge. The one-sidedness of a Möbius strip only exists if it is not cut. If it is, the strip becomes two sided. One might say that the one-sidedness is an emergent global property of the complete Möbius strip.

The higher up one goes in a given hierarchy of emergent behavior, the more the organization seems completely independent of the rules determining the behavior of the levels below — which, nevertheless, is not to deny that the higher-order rules are in some sense inherently determined by the properties of the component parts.[12] But it is the definition of "inherently determined" that contains the essence of the problem.

How can one resolve this conundrum? The answer may lie in the new, internal degrees of freedom that appear as one ascends a hierarchy of emergence. Consider first a simple example from elementary classical mechanics that has relevance to the formation of molecules and hence also to biology. The number of positional degrees of freedom for $2N$ particles is given by the product of the number of particles and the number of coordinates needed to specify the location of each of the particles in 3-dimensional space.

---

[11]Holden, A. V. (ed.), *Chaos* (Princeton University Press, Princeton, 1986).
[12]Mayr's discussion of "Explanatory Reductionism" is relevant here.

This is $2N \times 3 = 6N$. Now if the particles are combined so as to produce N bonded pairs, with some bonding distance associated with each pair, the number of external degrees of freedom for the pairs is reduced to $3N$. However, new degrees of freedom internal to each of the pairs have appeared — the distance between the particles constituting each pair, and the two angles needed to specify the orientation of each pair in three-dimensional space, a total of three internal degrees of freedom. Notice that the total number of degrees of freedom ($3N$ to locate the pairs in 3-space and $3N$ "emergent", internal degrees of freedom) has remained constant.

If we had started with $3N$ particles having $9N$ degrees of freedom and combined them to form N linear chain triplets, the original number of degrees of freedom would be reduced to $3N$ but the emergent internal degrees of freedom would add up to $6N$ so that the sum remains $9N$. For a linear chain configuration, the internal degrees of freedom are the two distances between the particles constituting each triplet, the two angles needed to orient the line between the first and second particle, and the two angles needed to orient the line from the second to third particles, a total of six for each triplet. For all the triplets, the total number of internal degrees of freedom would then be $6N$. Adding the $3N$ degrees of freedom needed to locate the N triplets in 3-space gives a grand total of $9N$. If the configuration of the triplets is changed to a triangular configuration, the definition of the internal degrees of freedom will also change: for each triplet the internal degrees of freedom would be the three distances between the particles comprising each triplet and the three angles needed to orient the triangular configuration in 3-space (two angles to orient the normal to the configuration and one angle for rotation about the normal).

The concept of degrees of freedom comes from classical mechanics where the number of degrees of freedom may be reduced because of the existence of constraints. If a system with n degrees of freedom having coordinates $q_1, q_2, \ldots, q_n$ is subject to k constraint equations having the form $f_r(q_1, q_2, \ldots q_n) = 0$, $r = 1, 2, \ldots k$, these equations can be solved for k of the coordinates in terms of the remaining $n-k$. The resulting $n-k$ coordinates may be varied independently without violating the constraints. The system thus has $n-k$ degrees of freedom and $q_1, q_2, \ldots, q_{n-k}$ may be taken as generalized coordinates, the constraints having been eliminated. An example might be helpful.

Consider a single particle free to move in three-dimensional space and thus having three degrees of freedom. Now constrain the particle to move on a surface given by $f(x, y, z) = 0$. This is the constraint equation, and it can be solved for z in terms of x and y, which become the generalized coordinates. The constrained particle now only has two degrees of freedom. The extra degree of freedom, unlike the composite particle examples given above, is eliminated by the constraint equation. Constraints could, of course, also exist for a composite particle in which case some of the internal degrees of freedom could be eliminated using the same procedure given above.

The conservation of the number of degrees of freedom is subtler in quantum mechanics. Take, for example, the case of the helium atom. A helium atom is comprised of a nucleus (considered as a single particle) and two electrons. As separate particles, assumed to be localized in space, the number of degrees of freedom for the three particles is nine. The combination of the three particles to form a helium atom would lead to three degrees of freedom for the location of the nucleus and six additional degrees of freedom consisting of the quantum numbers n, l, and m for each of the electrons. Of course, this is not the whole story since quantum mechanics sets additional constraints on the numerical values of the quantum numbers n, l, and m.

What one may call emergent behavior already appears even at the lowest level of reductionist exploration achieved to date. Ordinary matter derives its mass from the kinetic and potential energy of the massless gluons and nearly massless quarks that makeup protons, neutrons, and therefore all atomic nuclei. The theory (which can be viewed as a set of rules) governing their behavior is called quantum chromodynamics (QCD). Unlike many other theories, QCD has no adjustable parameters other than its overall coupling strength. Nonetheless, it has not proved possible to describe nuclei in terms of their fundamental quark and gluon constituents. Instead, so-called "effective" models of nuclei have been developed: in high-energy physics, nuclei are treated as a collection of free quarks; in the low-energy regime, on the other hand, one uses the many-body and shell models.

As one lowers the energy scale to the point where the interaction between distinct protons and neutrons becomes important, new internal degrees of freedom appear that are not contained in the higher energy theory. One might well view the properties of the low-energy regime as being emergent. Although, it is still not clear whether one should consider this a case of weak or strong emergence.

Similarly, the structure and variety of all atoms are determined by the rules of quantum mechanics. But the form of the lattice they or their compounds form may depend on emergent degrees of freedom such as temperature and pressure reflecting environmental factors. Although one might argue the varieties of structural forms depending on such environmental factors are emergent, it is also possible to argue — at least in principle — that the underlying quantum mechanical rules could take them into account.

An example where this is not possible is the chemistry of saturated hydrocarbons. The rules of quantum mechanics certainly determine the bonding of carbon and hydrogen, and no matter how structurally complex the hydrocarbon, these rules are faithfully obeyed. But the rules of quantum mechanics say nothing about how many carbon atoms may form a chain or whether they form straight chains or branched-chain carbon skeletons. There are emergent degrees of freedom that appear when atoms combine to form these hydrocarbon molecules. It is these emergent degrees of freedom that determine the chemical properties of the saturated hydrocarbons and these chemical properties could well be viewed as an emergent property of a complex system (saturated hydrocarbon molecules) not fully determined by the underlying quantum mechanical rules governing the bonding of hydrogen and carbon atoms.

The emergent rules that govern the chemistry of saturated hydro-carbons are *dependent* on the underlying rules governing the bonding of hydrogen and carbon, but are not *determined* by these rules. That is, the emergent rules cannot be derived from the underlying quantum mechanical rules governing hydrogen-carbon bonding. The difference here is similar to that found in mathematics between *necessity* and *sufficiency*. It is this distinction that should be used to inform the definition of reductionism, particularly in biology with its hierarchical organization, in light of the reality of emergent phenomena.

In the same way, the rules of chemical bonding (again reflecting the rules of quantum mechanics) specify the structure of DNA, but not the sequence of bases. The possible sequences of bases and length of the DNA molecule itself again constitute emergent degrees of freedom not specified by the rules of chemical bonding. The sequence of bases determines the genes coding for proteins, small RNAs, etc., and it is roughly at this level that the environment begins to play a significant role in the evolution of life through the Darwinian process of variation and selection.[13] But it is not only the set of genes that is responsible for the diversity of animal forms. Of primary importance are differences in gene regulation during ontogeny.[14]

The sequence of bases in DNA contains regulatory code that governs gene expression both in time and location. While this code constitutes another higher-level set of rules, they are rules that have the additional property of being able to change with time in response to environmental selection at the organismal level. The emergent number of degrees of freedom appearing at this level vastly exceeds those at lower hierarchic levels. The whole issue of epigenetics — defined as heritable changes in gene expression not due to changes in base sequence, essentially what allows cells having the same genetic inheritance to make up the variety of cell types comprising an organism — and its role in evolution is still an active area of research.[15,16]

---

[13] In terms of the origin of life, molecular evolution and the earliest living creatures were of course also subject to Darwinian variation and selection.

[14] Pennisi, E., "Searching the genome's second code", *Science* **306** (2004), 632–635.

[15] *Epigenetics*, Nature Insight: *Nature* **447** (2007), 395–440.

[16] Jablonka, E. and Lamb, M.J., *Evolution in Four Dimensions: Genetic, Epigenetic, Behavioral, and Symbolic Variation in the History of Life* (The MIT Press, Cambridge, 2006).

At an even higher level, while DNA surely determines the structure of living creatures, it would be impossible to derive their social behavior and organization from only the sequence of bases in DNA.

The hierarchy above, starting from the lowest reductionist level now known has led to the rules governing ontogeny. Indeed, from an Olympian point of view, life itself may be viewed as an emergent property of matter. But the hierarchy does not apparently end and may well be truly open in the sense of Köestler.[17]

Consciousness and intelligence appear to emerge gradually as the complexity of life increases. Simultaneously, and as a parallel development, a social structure comes into existence. Social behavior can be as simple as that of slime molds when forming a fruiting body, be relatively complex as in the behavior of an ant colony, or be represented by the far more complex behavior of human societies. All appear as forms of emergent behavior.

If the idea that emergent behavior results from the coming into being of new, internal degrees of freedom that arise as one ascends a given hierarchy of emergence is to hold, the inverse should also be true in the sense that a reductionist analysis should eliminate degrees of freedom in the process of descending the hierarchy through reductionist analysis. Here, reductionism is defined as gaining an understanding of a complex system through detailed analysis of the components of the system and their interactions. From the examples of emergence given above, this would seem to be almost trivially true.

In sum, one should view emergence and reductionism as opposite sides of the same coin. Dissecting complex behavior from the top down eliminates internal degrees of freedom in the course of analysis, while emergent phenomena occur when internal degrees of freedom appear when combining component elements into more complex systems. If individual ants are studied to determine their rules of interaction, there is nothing mysterious about the process. But given those rules, one cannot predict the behavior of the colony because

---

[17]Köestler, A., *The Ghost in the Machine* (The Macmillan Co., New York, 1968).

the new degrees of freedom that appear in the collective colony cannot be deduced from the rules of interaction — these rules are *necessary* but not *sufficient* to predict the emergent behavior. It is the unexpected consequences of the additional degrees of freedom that appear mysterious.

www.ingramcontent.com/pod-product-compliance
Lightning Source LLC
Chambersburg PA
CBHW050630190326
41458CB00008B/2210